Differential Equations

ACADEMIC PRESS RAPID MANUSCRIPT REPRODUCTION

Proceedings of the Eighth Fall Conference on Differential Equations
Held at Oklahoma State University, October 1979

DIFFERENTIAL EQUATIONS

edited by

SHAIR AHMAD
MARVIN KEENER

Department of Mathematics
Oklahoma State University
Stillwater, Oklahoma

A. C. LAZER

Department of Mathematics
University of Cincinnati
Cincinnati, Ohio

ACADEMIC PRESS 1980
A SUBSIDIARY OF HARCOURT BRACE JOVANOVICH, PUBLISHERS
New York London Toronto Sydney San Francisco

ACADEMIC PRESS, INC.
111 Fifth Avenue, New York, New York 10003

United Kingdom Edition published by
ACADEMIC PRESS, INC. (LONDON) LTD.
24/28 Oval Road, London NW1 7DX

Library of Congress Cataloging in Publication Data

Fall Conference on Differential Equations, 8th, Oklahoma State University, 1979.
Differential equations.

1. Differential equations—Congresses. I. Ahmad, Shair. II. Keener, Marvin. III. Lazer, A.C. IV. Title.
QA370.F34 1979 515.3′5 80-16549
ISBN 0-12-045550-1

PRINTED IN THE UNITED STATES OF AMERICA

80 81 82 83 9 8 7 6 5 4 3 2 1

CONTENTS

CONTRIBUTORS

Numbers in parentheses indicate the pages on which authors' contributions begin.

Shair Ahmad (103), *Department of Mathematics, Oklahoma State University, Stillwater, Oklahoma 74078*

Prem N. Bajaj (115), *Department of Mathematics, Wichita State University, Wichita, Kansas 07208*

Peter W. Bates (123), *Department of Mathematics, Texas A&M University, College Station, Texas 77843*

S. R. Bernfeld (127), *Department of Mathematics, University of Texas at Arlington, Arlington, Texas 76010*

G. J. Butler (135), *Department of Mathematics, University of Alberta, Edmonton, Alberta, Canada T6G 2G1*

Alfonso Castro (149), *C.I.E.A. del I.P.N., México 14, D.F. Mexico 14740*

Lamberto Cesari (1), *Department of Mathematics, University of Michigan, Ann Arbor, Michigan 48104*

D. L. DeAngelis (271), *Oak Ridge National Laboratory, Oak Ridge, Tennessee 37830*

Jerome Eisenfeld (161), *Department of Mathematics, University of Texas at Arlington, Arlington, Texas 76010*

Ivar Ekeland (123), *Départment de Mathematiques, Université Paris-Dauphine, Paris, France*

Jack K. Hale (23), *Department of Mathematics, Brown University, Providence, Rhode Island 02912*

James N. Hanson (171), *Computer Science Department, Cleveland State University, Cleveland, Ohio 44115*

T. L. Herdman (187), *Department of Mathematics, Virginia Polytechnic Institute and State University, Blacksburg, Virginia 24061*

Lloyd K. Jackson (31), *Department of Mathematics, University of Nebraska, Lincoln, Nebraska 68588*

Ronald Knight (193), *Department of Mathematics, Northeast Missouri State University, Kirksville, Missouri 63501*

Alan C. Lazer (199), *Department of Mathematics, University of Cincinnati, Cincinnati, Ohio 45221*

Ann L. LeFever (221), *Department of Biology, Marquette University, Milwaukee, Wisconsin 53233*

Roger C. McCann (215), *Department of Mathematics, Mississippi State University, Mississippi State, Mississippi 39762*

P. J. McKenna (199), *Department of Mathematics, University of Florida, Gainesville, Florida 32611*

Stephen J. Merrill (221), *Department of Mathematics and Statistics, Marquette University, Milwaukee, Wisconsin 53233*

A. N. Michel (51), *Department of Electrical Engineering and the Engineering Research Institute, Iowa State University, Ames, Iowa 50010*

R. K. Miller (51), *Department of Mathematics, Iowa State University, Ames, Iowa 50010*

R. Kent Nagle (235), *Department of Mathematics, University of South Florida, Tampa, Florida 33620*

W. M. Post (271), *Oak Ridge National Laboratory, Oak Ridge, Tennessee 37830*

Jorge Salazar (103), *Colegio Universitario de Caracas, Caracas, Venezuela*

L. Salvadori (127), *Dipartimento di Matematica, Universitá di Trento, Italy*

Klaus Schmitt (65), *Department of Mathematics, University of Utah, Salt Lake City, Utah 84112*

Peter Seibert (249), *Universidad Centro Occidental, Barquisimeto, Estado Lara, Republica de Venezuela, Venezuela*

George R. Sell (87), *Department of Mathematics, Univeristy of Minnesota, Minneapolis, Minnesota 55455*

Karen Singkofer (235), *University of Southern California, Los Angeles, California 90007*

S. J. Skar (51), *Department of Mathematics, Iowa State University, Ames, Iowa 50010*

C. C. Travis (271), *Oak Ridge National Laboratory, Oak Ridge, Tennessee 37830*

PREFACE

The Eighth Fall Conference on Differential Equations, which was held at Oklahoma State University in October 1979, is the continuation of the Seventh Midwest Conference on Differential Equations, which was held at the University of Missouri—Rolla, the year before.

The First Midwest Conference on Differential Equations was initiated by Professor Stephen R. Bernfeld, then at the University of Missouri—Rolla, and Professor Paul E. Waltman of the University of Iowa. It was held at the University of Iowa during fall of 1972. The other such conferences were held at the University of Missouri—Columbia, University of Nebraska, Northern Illinois University, Southern Illinois University, and Iowa State University.

The most recent conference, which was held at Oklahoma State University, was somewhat broader in scope in the sense that no special emphasis was placed on any particular area of differential equations. There was equal emphasis on partial differential equations, while in the past the main emphasis had been on ordinary differential equations. There was also no special emphasis placed on the geographic locations of the participants. For this reason, we felt that it would be more appropriate to call it the Eighth Fall Conference on Differential Equations instead of the Eighth Midwest Conference on Differential Equations.

The editors wish to express their appreciation and thanks to Ruth Duncan for her excellent work in the preparation of the manuscript.

SHAIR AHMAD

HYPERBOLIC PROBLEMS: EXISTENCE AND APPLICATIONS

Lamberto Cesari

University of Michigan

INTRODUCTION

We consider here the question of existence of solutions of abstract operator equations of the form

$$Ex = Nx, \quad x \in X,$$

where X is a Hilbert space, E is the linear operator with a possibly infinite dimensional kernel X_0, and E is such that the partial inverse H of E on the quotient space X/X_0 is bounded, but not necessarily compact. Our theorems therefore apply to quasilinear hyperbolic partial differential equations and systems, in particular wave equations.

Our purpose here is to point out how much of the recent developments in the theory of nonlinear elliptic partial differential equations can be extended naturally to obtain existence of solutions of nonlinear hyperbolic problems.

In the recent years there has been an extensive literature on the question of existence of solutions to quasilinear elliptic equations of the type $Ex = Nx$ (E being a linear operator with a finite dimensional kernel, the partial inverse of E being compact, and N nonlinear), and we have already shown [2,3,4,5,7] that in this situation suitably conceived

ISBN 0-12-045550-1

abstract existence theorems essentially contain most of the results just mentioned for elliptic problems.

It is therefore the purpose of this paper to show that analogous abstract existence theorems can be obtained in the present more general situation (X_o infinite dimensional and H bounded but not compact). In recent papers L. Cesari and R. Kannan [9,10] have developed this program showing, as in the elliptic case, that most known specific results for the hyperbolic case, and new simple criteria, can be derived from our abstract theorems. In particular, existence statements for specific problems of the forms $u_{tt} - Au = f(.,.,u)$, or $u_{tt} - Au = f(.,.,u,u_t)$ can be derived, which contain as particular cases results which had been proved by W. S. Hall [13,14] and H. Petzeltova [17] only for $f = \varepsilon g$, ε a small parameter.

The basic formulation of our abstract theorems for the hyperbolic case is about the same as in the elliptic case. It appears therefore that some unification has been brought about in a rather large variety of specific situations.

2. <u>The Auxiliary and Bifurcation Equations</u>. Let X and Y be real Banach spaces and let $\|x\|_X$, $\|y\|_Y$ denote the norms in X and Y respectively. Let $\mathcal{D}(E)$ and $\mathcal{R}(E)$ be the domain and range of the linear operator $E : \mathcal{D}(E) \to Y$, $\mathcal{D}(E) \subset X$ and let $N : X \to Y$ be an operator not necessarily linear. We shall now consider the equation

$$Ex = Nx, \quad x \in \mathcal{D}(E) \subset X. \qquad (1)$$

Let $P : X \to X$, $Q : Y \to Y$ be projection operators (i.e., linear bounded, and idempotent), with ranges and null spaces given by

$$\mathcal{R}(P) = PX = X_o \ , \quad \ker \ P = \mathcal{R}(I - P) = (I - P)X = X_1 \ ,$$

$$\mathcal{R}(Q) = QY = Y_o \ , \quad \ker \ Q = \mathcal{R}(I - Q) = (I - Q)Y = Y_1 \ .$$

We assume that P and Q can be so chosen that $\ker E = X_o = PX$, $\mathcal{R}(E) = Y_1 = (I - Q)Y$. This requires that $\ker E$ and $\mathcal{R}(E)$ are closed in the topologies of X and Y respectively. Then, E as a linear operator from $\mathcal{D}(E) \cap X_1$ into Y_1 is one-to-one and onto, so that the partial inverse $H : Y_1 \to \mathcal{D}(E) \cap X_1$ exists as a linear operator. We assume that H is a bounded linear operator, not necessarily compact, and that the following axioms hold:

i) $H(I - Q)E = I - P$,

ii) $EP = QE$,

iii) $EH(I - Q) = I - Q$.

We have depicted a situation which is rather typical for a large class of differential systems, not necessarily self-adjoint. Let L be a constant such that $\|Hy\|_X \le L\|y\|_Y$ for all $y \in Y_1$. We have seen in [2] that equation (1) is equivalent to the system of auxiliary and bifurcation equations

$$x = Px + H(I - Q)Nx \ , \tag{2}$$

$$Q(Ex - Hx) = 0 \ . \tag{3}$$

If $x^* = Px \in X_o$, and $\ker E = X_o = PX$, then these equations become

$$x = x^* + H(I - Q)Nx \ , \tag{4}$$

$$QNx = 0 \ , \quad \text{with} \quad x = x^* + x_1, \ x^* \in X_o, \ x_1 \in X_1 \ . \tag{5}$$

Thus, for any $x^* \in X_o$, the auxiliary equation (4) has the form of a fixed point problem $x = Tx$, with $Tx = x^* + H(I - Q)Nx$.

For $X = Y$, a real Hilbert space, Cesari and Kannan [7] have given sufficient conditions for the solvability of equation (1) in terms of monotone operator theory.

Again, for $X = Y$ a real separable Hilbert space, E self-adjoint and N Lipschitzian, Cesari (cf [2]; § 1, nos. 3-5) has shown that it is always possible to choose X_o and hence, P, Y, Q in such a way that the operator T is a contraction map in the norm of X, and thus, the auxiliary equation $x = Tx$ has a unique solution in a suitable ball in X by the Banach fixed point theorem. McKenna [16] has extended the result to nonself-adjoint operators E in terms of its dual E^* and the self-adjoint operators EE^* and E^*E. Recently Cesari and McKenna [11] have indicated a set-theoretic basis for the extension of the basic arguments to rather general situations.

In general for E, not necessarily self-adjoint and X, Y real Banach spaces as stated above, Cesari and Kannan in a series of papers have considered the situation where Y is a space of linear operators on X, so that the operation (y, x), $Y \times X \to R$ is defined, is linear in both x and y, under the following natural assumption

$$(\pi_1) : |(y, x)| \leq K \|y\|_Y \|x\|_X$$

for some constant K and all $x \in X$, $y \in Y$.

We can always choose norms in X or Y, or the operation (y, x), in such a way that $K = 1$. Furthermore, we assume that: (π_2) for $y \in Y$, we have $y \in R(E) = Y_1$, i.e., $Qy = 0$ if and only if $(Qy, x^*) = 0$ for all $x^* \in X_o$.

As simple examples of the above situation, we have the following. Here G denotes a bounded domain in any t-space R^ν, $t = (t_o, \ldots, t_\nu)$, $\nu \geq 1$.

a) $X = Y = L_2(G)$, $|(y, x)| = |\int_G y(t)x(t)dt| \leq \|y\| \|x\|$

with usual norms in L_2.

b) $X = L_2(G)$ with L_2 norm $\|x\|$, $Y = L_\infty(G)$ with norm $\|y\|_\infty$, and then

$$|(y, x)| = |(\text{meas } G)^{-\frac{1}{2}} \int_G y(t)x(t)dt| \leq \|y\|_\infty \|x\| .$$

c) $X = L_\infty(G)$ with usual norm $\|x\|_\infty$, $Y = L_\infty(G)$ with norm $\|y\|_\infty$, and then again

$$|(y, x)| = |(\text{meas } G)^{-1} \int_G y(t)x(t)dt| < \|y\|_\infty \|x\|_\infty .$$

d) $X = H^m(G)$ with usual Sobolov norm $\|x\|_m$, $Y = L_2(G)$, and then

$$|(y, x)| = |\int_G y(t)x(t)dt| \leq \|y\| \|x\| \leq \|y\| \|x\|_m .$$

3. <u>An Abstract Theorem for the Elliptic Case</u>. Let X, Y be real Banach spaces, Y a space of linear operators on X with linear operation (y, x) satisfying requirements (π_1) and (π_2). Let us consider equation (1) in X with $X_o = \ker E$ of finite dimension, and let H be compact, and P, Q as in § 2.

Let $w = (w_1, w_2, \ldots, w_m)$ be an arbitrary basis for the finite dimensional space $X_o = \ker E = PX$, $1 \leq m = \dim \ker E < \infty$. For $x^* \in X_o$ we have $x^* = \Sigma_i c_i w_i$, or briefly $x^* = cw$, $c = (c_1, \ldots, c_m) \in R^m$ and there are constants $0 < \gamma' \leq \gamma < \infty$, such that $\gamma'|c| \leq c\|w\| \leq \gamma|c|$, where $| \ |$ is the Euclidean norm in R^m. The coupled system of equations can now be written in the form

$w = cw + H(I - Q)Nx$ and $(QNX, w) = 0$. Let $L = \|H\|$, $\chi = \|Q\|$, $\chi' = \|I - Q\|$. The following existence theorem holds:

Theorem (3.i). Let X, Y be real Banach spaces and let E, H, P, Q be as in Section 2. Let $X_o = \ker E$ be nontrivial and finite dimensional, and H linear, bounded and compact. If (a) there is a constant $J_o > 0$ such that $\|Nx\| \le J_o$ for all $x \in X$, and (b) there is a constant $R_o \ge 0$ such that $(QNx, x^*) \le 0$ [or $(QNx, x^*) \ge 0$] for all $x \in X$, $x^* \in X_o$ with $Px = x^*$, $\|x^*\| \ge R_o$, $\|x - x^*\| \le L\chi' J_o$, then the equation $Ex = Nx$ has at least a solution $x \in \mathcal{D}(E) \cap X$.

For a proof of this theorem, we refer to Cesari and Kannan [8] and subsequent papers by Cesari [3, 4, 5]. A topological proof may be seen in Kannan and McKenna [15] for the case $X = Y$, $P = Q$, X real Hilbert space. For extensions of (3.i) to operators N with growth $\|Nx\| \le J_o + J_1 \|x\|^k$, $0 < k < 1$, or arbitrary growth $\|Nx\| \le \Phi(\|x\|)$, including the former case with $1 \le \kappa < \infty$, we refer to Cesari [3,4].

Well known results of A. C. Lazer and D. E. Leach, E. M. Landesman and A. C. Lazer, S. A. Williams, and D. G. de Figueiredo for elliptic problems can be shown to be particular cases of the abstract theorem above and variants (cf L. Cesari [2] for references and proofs).

4. Preliminary Considerations Concerning the Hyperbolic Case. Let E, N be operator from their domains $\mathcal{R}(E)$, $\mathcal{R}(N)$ in $\mathcal{X}$ to $\mathcal{Y}$, both $\mathcal{X}$ and $\mathcal{Y}$ real Banach or Hilbert spaces, and let us consider the operator equation

$$Ex = Nx$$

as in Section 2. Its solutions x in $\mathcal{X}$ may be expected to be usual solutions, or generalized solutions according to the choice of $\mathcal{X}$. We shall consider first smaller spaces X and Y, say $X \subset \mathcal{X}$, $Y \subset \mathcal{Y}$, both real Hilbert spaces, and we shall assume that the inclusion map $j : X \to \mathcal{X}$ is compact.

We shall then construct a sequence of elements $[x_k]$, $x_k \in X$, which is bounded in X, or $\|x_k\|_X \leq M$. Then, there is a subsequence, say still [k] for the sake of simplicity, such that $[jx_k]$ converges strongly in $\mathcal{X}$. On the other hand, X is Hilbert, hence reflexive, and we can take the subsequence, say still [k], in such a way that $x_k \to x$ weakly in X. Actually, $\xi = jx$, that is, ξ is the same element $x \in X$ thought of as an element of $\mathcal{X}$, in other words:

<u>Theorem (4.i)</u> If $x_k \to x$ weakly in X and $jx_k \to \xi$ strongly in $\mathcal{X}$, then $\xi = jx$.

Indeed, $j : X \to \mathcal{X}$ is a linear compact map, hence continuous (see, e.g., [1], p. 285, Th.17.1). As a consequence, $x_k \to x$ weakly in X implies that $jx_k \to jx$ weakly in $\mathcal{X}$ (see, e.g., [1], p.295, pr. no.12). Since $jx_k \to \xi$ strongly in $\mathcal{X}$, we have $\xi = jx$.

We shall assume that X_1 and x_o contain finite dimensional subspaces X_{1n}, X_{on} such that $X_{1n} \subset X_{1,n+1} \subset X_1$, $X_{on} \subset X_{o,n+1} \subset X_o$, $n = 1,2,\ldots$, with $\cup_n X_{1n} = X_1$, $\cup_n X_{on} = X_o$, and assume that there are projection operators $R_n : X_1 \to X_1$, $S_n : X_o \to X_o$ with $R_n X_1 = X_{1n}$, $S_n X_o = X_{on}$ (cf for similar assumptions E. H. Rothe [18]). Since X is a real Hilbert space, we may think of R_n and S_n as orthogonal projections, and then $\|R_n x\|_X \leq \|x\|_X$, $\|S_n x^*\| \leq \|x^*\|_X$ for all $x \in X_1$ and $x^* \in X_o$.

Thus, we see that in the process of limit just mentioned, $x_k \to x$ weakly in X, $jx_k \to jx$ strongly in $\mathcal{X}$, the limit element can still be thought of as belonging to the smaller space X. This situation is well known in the important case $X = W_2^N(G)$, $\mathcal{X} = W_2^n(G)$, $0 \le n < N$, $X \subset \mathcal{X}$, G an open set in some R^ν, $\nu \ge 1$. Then, the weak convergence $x_k \to x$ in $W_2^N(G)$ implies the strong convergence $jx_k \to jx$ in $W_2^n(G)$, and $\xi = jx$ is still an element of the smaller space X.

Concerning the subspaces X_{on} of X_o it is not restrictive to assume that there is a complete orthonormal system $[\nu_1, \nu_2, \dots, \nu_n, \dots]$ in X_o and that $X_{on} = sp\,(\nu_1, \nu_2, \dots, \nu_n)$, $n = 1, 2, \dots$. We shall further assume that there is a complete orthonormal system $(\omega_1, \omega_2, \dots, \omega_n, \dots)$ in Y, such that $(\omega_i, \nu_i) \neq 0$ and $(\omega_i, \nu_j) = 0$ for all $i \neq j$. We shall take $Y_{on} = sp(\omega_1, \dots, \omega_n)$ and denote by S_n' the orthogonal projection of Y_o onto Y_{on}. Then, $S_n'QNx = 0$ if and only if $(QNx, \nu_j) = o$, $j = 1, \dots, n$, and this holds for all $n = 1, 2, \dots$.

We consider now the coupled system of operator equations

$$x = S_nPx + R_nH(I - Q)Nx \,, \tag{6}$$

$$0 = S_n'QNx \,. \tag{7}$$

We note that we have $S_n'QNx = 0$ if and only if $(QNx, x^*) = 0$ for all $x^* \in X_{on}$.

We shall now define a map $\alpha_n : Y_{on} \to X_{on}$ by taking

$$\alpha_n y = \textstyle\sum_1^n (y, \omega_i)\nu_i \,.$$

Then, we have $0 = S_n'QNx$ if and only if $0 = \alpha_n S_n' QNx$. We conclude that system (6, 7) is equivalent to system

$$x = S_n Px + R_n H\,(I - Q)Nx \,, \tag{8}$$

$$0 = \alpha_n S_n' QNx \,. \tag{9}$$

(4.ii) (a lemma) Under the hypotheses above, let us assume that there are constants $R, r > 0$ such that

i) for all $x^* \in X_o$, $x_1 \in X_1$, $\|x^*\| \le R$, $\|x_1\| \le r$, we have $\|N(x^* + x_1)\|_y \le L^{-1}r$;

ii) for all $\|x^*\| = R_o$, $\|x_1\| \le r$ we have $(QN(x^* + x_1) , x^*) \ge 0$ [or ≤ 0] .

Then, for every n, system (6), (7) has at least a solution $x_n \in \mathcal{D}(E) \cap (X_{on} \times X_{1n}) = x^*_{on} + x_{1n}$, $S_n Px_n = x^*_{on}$ with $\|x_n\| \le M$ and M independent of n .

<u>Proof</u>. If we consider the subset C_n of $X_{on} \times X_{1n}$ made up of all $x = x_{on} + x_{1n}$ with $\|x_{on}\| \le R_o$, $\|x_{1n}\| \le r$, we see that

$$\|R_n\, H(I - Q)Nx\| \le r \quad \text{for all} \quad x \in C_n \,,$$

$$(\alpha_n S_n' QNx\, , x^*) \ge 0 \quad [\text{or} \le 0] \quad \text{for all } x \in C_n \text{ with } x^* = R.$$

In other words, the assumptions actually used in the proofs in [8] and in [15], are satisfied with $L^{-1}r$ replacing J_o and the same R_o .

Now the compactness of the bounded operator R_nH follows from the fact that R_nH has a finite dimensional range, and the finite dimensionality of the kernel of E is now replaced by the fact that the range of $\alpha_n S_n QN$ is certainly finite dimensional. The proofs of the aforementioned theorems repeat now verbatim, with J_o replaced by $L^{-1}r$. The bound $M = (R_o^2 + L^2 J_o^2)^{\frac{1}{2}}$ is now replaced by the bound $M = (R_o^2 + r^2)^{\frac{1}{2}}$ certainly independent of n.

5. <u>An Abstract Theorem for the Hyperbolic Case</u>. In order to solve the equation $Ex = Nx$ we now adopt a "passage to the limit argument." We assume that both the Hilbert spaces X and Y are contained in real Banach (or Hilbert) spaces $\mathcal{X}$ and $\mathcal{Y}$ with compact injections $j : X \to \mathcal{X}$, $j' : Y \to \mathcal{Y}$. Actually, we can limit ourselves to the consideration of the spaces $\overline{\mathcal{X}}$ and $\overline{\mathcal{Y}}$, made up of limit elements from sequences in X and Y respectively as mentioned in Section 4. Hence, $\overline{\mathcal{X}}$ is identical to X, and $\overline{\mathcal{Y}}$ is identical to Y, though they may have different topologies. We shall write $\overline{\mathcal{X}} = jX$, $\overline{\mathcal{Y}} = jY$.

Analogously, we take $\mathcal{X}_o = jX_o$, $\mathcal{Y}_o = j'Y_o$, $\mathcal{X}_1 = jX_1$, $\mathcal{Y}_1 = j'Y_1$, and the linear operators $\mathcal{P} : \mathcal{X} \to \mathcal{X}_o$, $\mathcal{Q} : \overline{\mathcal{Y}} \to \mathcal{Y}_o$ are then defined by $\mathcal{P}x = x^*$ in $\overline{\mathcal{X}}$ if $Px = x_o$ in X; $\mathcal{Q}y = y^*$ in $\mathcal{Y}$ if $Qy = y^*$ in Y.

We now assume the following:

(C) $x_n \to x$ weakly in X and $x_n \to x$ strongly in $\mathcal{X}$, implies that $Nx_n \to Nx$ strongly in $\mathcal{X}$, $S_n P x_n \to \mathcal{P}x$ strongly in $\mathcal{X}$, and $R_n x_n \to x$ strongly in $\mathcal{X}$.

By the lemma, we have elements $x_n \in X_n$ such that

$$x_n = S_n P_n x_n + R_n H (I - Q) N x_n, \tag{10}$$

$$0 = \alpha_n S_n' Q N x_n, \tag{11}$$

where $\|x_n\| \le M$ for all n. Hence, there exists a

subsequence, say still $[x_n]$, such that $x_n \to x$ weakly in X and $x_n \to x$ strongly in $\mathcal{X}$. Then, by (10, 11), proceeding to the limit, we have

$$x = Px + H(I - Q)Nx, \; 0 = QNx, \; x \in \bar{X} .$$

Indeed, as $n \to \infty$, S_n converges to the identity $Y_o \to Y_o$ and α_n converges to a homeomorphism $\alpha : Y_0 \to X_0$ in the sense that $S_n y \to y$, $\alpha_n y \to \alpha y$ as $n \to \infty$

We now remark that, in $\bar{X}$ the operator E may have no meaning and thus, the concept of solution of $Ex = Nx$ has to be properly understood. However, $x \in \mathcal{X}$, and thus, by Section 4, x is still an element of X on which E is defined. Further, as a consequence of the hypotheses on P and H, we have $QE = EP = 0$ and $EH(I - Q) = I - Q$. Thus, from the above limit equation we have

$$\begin{aligned} Ex &= EPx + EH(I - Q)Nx + QNx \\ &= EPx + (I - Q)Nx + QNx = Nx . \end{aligned}$$

We summarize now the hypotheses and the conclusions concerning the operator equation $Ex = Nx$, we have obtained.

Theorem (5.i). Let $E : \mathcal{D}(E) \to Y \subset \mathcal{Y}$, $\mathcal{D}(E) \subset X \subset \mathcal{X}$, E a linear operator, $N : X \to Y$ a not necessarily linear operator, X, Y real Hilbert spaces, $\mathcal{X}, \mathcal{Y}$ real Banach or Hilbert spaces with compact injections $j : X \to \mathcal{X}$, $j : Y \to \mathcal{Y}$, with projection operators $P : X \to X$, $Q : Y \to Y$ and decompositions $X = X_o + X_1$, $Y = Y_o + Y_1$, $X_o = PX = \ker E$, $Y_1 = (I - Q)Y = \text{Range } E$, X_o infinitely dimensional, and bounded partial inverse $H : Y_1 \to X_1$, P, Q, H, E satisfying i), ii), iii) of Section 2. Let $\mathcal{Y}$ be a space of linear operators (y, x), or $\mathcal{Y} \times X \to$ Reals, satisfying (π_1), (π_2) of Section 2. Let X_{on}, X_{1n}, Y_{on} be finite dimensional subspaces of X_o, X_1, Y_o with orthogonal projection operators $R_n : X_1 \to X_1$, $S_n : X_o \to X_o$, $S_n' : Y_o \to Y_o$ with $R_n X_1 = X_{1n}$, $S_n X_o = X_{on}$, $S_n' Y_o = Y_{on}$, satisfying i) and ii) of (4.i), the other requirements in Section 4, and requirement (C) of the present Section. Then the equation $Ex = Nx$ has at least a solution (L. Cesari and R. Kannan [9]).

In [9,10] analogous theorems have been also proved under different conditions. In particular, the case of slow growth $\|Nx\| \le J_o + J_1 \|x\|^\nu$, $0 < \nu < 1$, and of arbitrary growth $\|Nx\| \le \Phi(\|x\|)$ have been investigated.

6. <u>Some Applications</u>.

(a) Let us consider the problem

$$
\begin{aligned}
&u_{tt} + u_{xxxx} = f(t,x,u,u_t) , \\
&u(t,0) = u(t,\pi) = u_{xx}(t,0) = u_{xx}(t,\pi) = 0, \qquad (12) \\
&u(t + 2\pi,x) = u(t,x), \quad -\infty < t < +\infty, \quad 0 < x < \pi .
\end{aligned}
$$

For $Eu = u_{tt} + u_{xxxx}$ with the conditions stated, then ker E has infinite dimension, and contains, in particular, all functions as $\sin k x \cos k^2 t$, $\sin k x \sin k^2 t$, $k = 1, 2, \ldots$. By the use of Theorem (5.i), theorems of the Landesman-Lazer type for problem (12) have been derived. In particular, if f is of the form $f = \varepsilon g$ with ε small, then the condition $g_{u_t} \ge \mu > 0$ is sufficient for the existence of periodic solutions. The latter is a result which was proved by Hana Petzeltova (1973) by specific arguments.

(b) Let us consider the problem

$$u_{tt} - u_{xx} = f(t,x,u) , \qquad (13)$$

or the more general one

$$u_{tt} + (-1)^p D_x^{2p} u = f(t,x,u) , \quad p \geq 1 ,$$

both with conditions of double periodicity

$$u(t + 2\pi, x) = u(t,x) = u(t, x + 2\pi) .$$

For $p = 1$, for instance, and $Eu = u_{tt} - u_{xx}$, then $\ker E$ is infinite dimensional and contains, in particular, all functions as $\sin kx \sin kt$, $\sin kx \cos kt$, $\cos kx \sin kt$, $\cos kx \cos kt$, $k = 0, 1, \ldots$. Actually, $\ker E = \{\phi(t - x) + \psi(t + x)\}$, ϕ, ψ 2π-periodic in R. By the use of Theorem (5.i), theorems of Landesman-Lazer kind for problem (13) have been derived. In particular, if f is of the form $f = \varepsilon g$ with ε small, then the condition $g_u \geq \mu > 0$ is sufficient for the existence of 2π-periodic solutions. The latter is a result which was proved by W. S. Hall (1967) by specific arguments.

(c) Let us consider the hyperbolic problem proposed by J. Mawhin and S. Fucik

$$\begin{aligned} &bu_t + u_{tt} - u_{xx} + g(u) = h(t,x) , \\ &u(t,0) = u(t,2\pi) = 0 , \\ &u(t + 2\pi,x) = u(t,x) , \end{aligned} \tag{14}$$

where $b \neq 0$ is a real constant, $g : R \to R$ is continuous and bounded, and h is 2π-periodic in t, and of class L^2 in the fundamental square G of the tx-plane. By the use of Theorem (5.i), the result of Mawhin and Fucik can be directly derived, namely, that for $b \neq 0$, $g(-\infty) < g(+\infty)$ both finite, and

$$g(-\infty) < (2\pi)^{-2} \int_0^{2\pi} \int_0^{2\pi} h(t,x)\,dtdx < g(+\infty) ,$$

problem (14) has at least a solution.

7. Specifics on the Wave Equation. The problem of the doubly periodic solutions of the wave equation

$$u_{xx} - u_{yy} = f(x,y,u) \ , \ (x,y) \in R^2 \ , \tag{15}$$

can be reduced to the problem of the doubly periodic solutions of the equation

$$u_{t\tau} = F(t,\tau,u) \tag{16}$$

and viceversa, by the usual transformations

$$x = t + \tau \ , \ y = t - \tau \ , \ t = 2^{-1}(x + y) , \ \tau = 2^{-1}(x - y) ,$$

$$U(t,\tau) = u(t+\tau, \ t - \tau) \ , \ F(t,\tau,u) = f(t + \tau, \ t - \tau \ , \ u) .$$

Let us concern ourselves here with problem (16) since the notations are easier. We shall write (16) in the usual form $Eu = Nu$ with the periodicity condition

$$u(t + T,\tau) = u(t,\tau) = u(t, \ \tau + T) \ .$$

Let $X = \{u(t,\tau)\}$ be the Banach space of all continuous and T-periodic functions $u(t,\tau)$ with norm $\|u\|_\infty$. Let $G = [0,T] \times [0,T]$. Then, for any $u \in x$,

$$u^*(t,\tau) = u(t,0) + u(0,\tau) - u(0,0) \ , \quad (t,\tau) \in R^2 \ , \tag{17}$$

is still an element of X which we will denote as "the boundary values of u" in the sense that $u = u^*$ on ∂G, and then naturally also on all straight lines

$$t = hT \ , \ t \in R \ , \quad \text{and} \quad t \in R \ , \ \tau = kT, \tag{18}$$

$$h,k = 0, \pm 1 , \pm 2 , \ldots .$$

Let us remark that every function $u^*(t,\tau) = u_0(t) + v_0(\tau) + u_{00}$, u_0, v_0 continuous and T-periodic on R, u_{00} constant, can be written in the form (17). For every $u \in X$ and u^* defined by (17), then $u_1 = u - u^*$ is zero on ∂G and on all straight lines (18).

Let X_0 be the subspace of all functions $u^* \in X$ of the form $u^* = u_0(t) + v_0(\tau) + u_{00}$, and let X_1 be the subspace of all $u_1 \in X$ which are zero on ∂G and on all straight lines (18). Let $P : X \to X$ denote the projection operator defined by $Pu = u^*$, where u^* is given by (17). Then $PX = X_0$, $(I - P)X = X_1$, $X = X_0 + X_1$, and every element $u \in X$ has a unique decomposition

$u = u^* + u_1$, $u^* \in X_0$, $u_1 \in X_1$, with

$$|u^*(P)| \le \|u\|_\infty \quad , \quad |u_1(P)| = |u(P) - u^*(P)| \le 2\|u\|_\infty \; .$$

It is easy to see that

$$\|P\| = 1 \quad , \quad \|I - P\| = 2 \; . \tag{19}$$

It is clear that $X_0 = \ker E$ is the null space of E, that is, the space of all elements $u \in X$ which are weak T-periodic solutions of $u_{t\tau} = 0$.

Let $Y = X$ with the same norm. If $F(t,\tau)$ is an element of Y, then F is continuous in R^2 and T-periodic in t and τ, and we take

$$\begin{aligned} \mu &= T^{-2} \int_0^T \int_0^T F(\xi, \tau)\,d\xi\,d\eta \, , \\ m(\tau) &= T^{-1} \int_0^T F(\xi, \tau)\,d\xi - \mu \, , \\ n(t) &= T^{-1} \int_0^T F(t, \eta)\,d\xi - \mu \, , \end{aligned} \tag{20}$$

$$F_0(t,\tau) = \mu + n(t) + m(\tau) \; .$$

Then, μ is a constant, and m, n are continuous T-periodic functions of R of mean value zero. We call μ, $m(t)$, $n(t)$ the system of the "mean values of F."

Let Y_0 denote the subspace of Y of all continuous and T-periodic functions of the form $F_0(t,\tau) = U(t) + V(\tau) + \mu$, U, V continuous T-periodic on R with mean value zero, and μ constant.

For each $F \in Y$ and F_0 defined by (20), then $F_1(t,\tau) = F(t,\tau) - F_0(t,\tau)$ is an element of Y whose mean values are all zero. Let Y_1 denote the subspace of Y of all $F_1 \in Y$ whose mean values are all zero. Let $Q : Y \to Y$ denote the projection operator defined by $QF = F_0$, where F_0 is given by (20). Then $QY = Y_0$, $(I - Q)Y = Y_1$, $Y = Y_0 + Y_1$, and every element $F \in Y$ has a unique decomposition $F = F_0 + F_1$, $F_0 \in Y_0$, $F_1 \in Y_1$. We have here

$$\begin{aligned} F_0(t,\tau) &= QF = m(\tau) + n(t) + \mu \\ &= T^{-1} \int_0^T F(\xi, \tau)\,d\xi + T^{-1} \int_0^T F(t,\eta)\,d\eta \\ &\qquad - T^{-2} \int_0^T \int_0^T F(\xi,\eta)\,d\xi\, d\eta \\ &= \tilde{m}(\tau) + \tilde{n}(t) - \mu , \end{aligned}$$

$$|\mu| \le \|F\|_\infty , \quad |m(\tau)| \le 2\|F\|_\infty , \quad |n(t)| \le 2\|F\|_\infty ,$$

$$|F_0(t,\tau)| \le 3\|F\|_\infty ,$$

and it is easy to see that

$$\|Q\| = 3 , \quad \|I - Q\| = 4 . \tag{21}$$

We can now define the operator $H : Y_1 \to X_1$, the partial inverse of E, by taking

$$v_1(t,\tau) = HF_1 = \int_0^t \int_0^\tau F_1(\xi,\eta)\,d\xi\, d\eta ,$$

or

$$v_1(t,\tau) = H(I - Q)F = \int_0^t \int_0^t [F(\xi,\eta) - m(\eta) - n(\xi) - \mu]\,d\xi d\eta$$

It is easy to see that $|v_1(t, \tau)| \le (T^2/4) \|F\|_\infty$, and more precisely that

$$\|H(I - Q)\| = T^2/4 . \qquad (22)$$

The problem of the T-periodic solutions of the equation

$$u_{t\tau} = F(t,\tau,u) \qquad (23)$$

is now reduced to the system

$$u_1 = H(I - Q)Nu \ , \quad u = u^* + u_1 \ , \quad \text{auxiliary equation,}$$

$$QNu = 0, \quad \text{or} \quad SQNu = 0 \ , \qquad \text{bifurcation equation,}$$

where $S : Y_o \to X_o$ is the identity, and $N : X \to Y$ is the operator $Nu = F(t,\tau,u(t,\tau))$.

We may consider now sets of the form

$$\Omega = S_o \times S_1 \ , \ S_o = \{u^* \in X_o \mid \|u^*\|_\infty \le R_o\} \ ,$$

$$S_1 = \{u_1 \in X_1 \mid \|u_1\|_\infty \le r\} \ ,$$

and the transformation $T : \Omega \to X$, or $(u^*, u_1) \to (\bar{u}^*, \bar{u}_1)$ defined by

$$T : \bar{u}_1 = K_1 u \ , \ \bar{u}^* = u^* - K_o u \ ,$$

$$u \in \Omega \ , \ u = u^* + u_1 \ , \ \bar{u} = \bar{u}^* + \bar{u}_1 \ , \qquad (24)$$

$$K_1 u = H(I - Q)Nu \ , \ K_o u = SQNu \ .$$

However, we shall introduce compactness properties, and to this effect we may, for instance, reduce ourselves to transformations $T_n : \Omega_n \to \Omega_n$ of certain subsets Ω_n into themselves, each Ω_n being obtained as the intersection of Ω with finite dimensional subspaces of C and further restrictions by means of seminorms. Then, the sequence $[u_n]$ of fixed elements $u_n = T_n u_n \in \Omega_n$ contains a subsequence which is uniformly convergent toward an element $u \in X$ which is a weak (continuous and T-periodic) solution of the problem $u_{t\tau} = F(t,\tau,u)$.

For instance, we have considered the problem of the existence of weak (continuous) solutions of the problem

$$u_{t\tau} = \phi(t,\tau) + g(u) \ , \tag{25}$$

$$u(t + T,\tau) = u(t,\tau) = u(t,\tau + T) \ .$$

Let us assume that

$$|\phi(P)| \leq c \ , \ |\phi(P) - \phi(Q)| \leq \lambda \ |P - Q| \ ,$$

$$|g(u)| \leq C \ , \ |g(u) - g(u)| \leq D \ |u - v| \ ,$$

for $|u|,|v| \leq R = R_o + r$,

$|g(u) - g(v)| \leq d|u - v|$ for $b \leq u, v \leq R$ and for $-R \leq u,v \leq -b$, $g(u) \geq B$ for $u \geq b$, $g(u) \leq -B$ for $u \leq -b$,

$|u - v - kg(u) + kg(v)| \leq \gamma \ |u - v|$ for $|u|,|v| \leq \nu$,

where $T, c, \lambda, C, B, b, D, k, \gamma, \nu$ are given constants. If these constants satisfy a certain system of inequalities, then there are numbers T_o , α_o , β_o , γ_o , δ_o such that, for $T \leq T_o$, $\alpha_o c + \lambda \leq \beta_o$, $\gamma_o c + \lambda \leq \delta_o$, problem (25) has at least a continuous T-periodic (weak) solution.

For instance, it was found that the continuous function $g(u)$, $u \in R$, defined by

$g(-u) = -g(u)$, $g(u) = \text{arc tan } u$ for

$0 \leq u \leq \nu = 1.3984 = \tan(o.95)$,

$g(u) = 0.95 + e \sin \sigma(u - \nu)$ for $u \geq \nu$,

where e , σ are arbitrary constants with $|e| \leq 0.06$, $|e\sigma| \leq 0.4$ satisfy the mentioned inequalities with

$D = 1$, $C = 0.95 + 0.06 = 1.01$, $b = \tan(0.89) = 1.2349$, $R_o = 3$, $R_1 = 1.61$, $r = 0.01$, $d = 0.4$, $\gamma = 0.66166$, $B = 0.95 - 0.06 = 0.89$, $T = 0.1$, and it is enough to require for ϕ that

$$(1.088)c + c' + \lambda \leq 0.1 \quad \text{and} \quad (1.34)c + c' + \lambda \leq 0.47234.$$

This example shows that, in our results, we do not require total monotonicity on the continuous function $g(u)$.

The considerations of the present Section are still in the lines of those of Sections 4 and 5 above, though based on operators initially discussed in Cesari ([6], 1965).

For details and proofs concerning the material discussed in Sections 1 - 7 of the present paper, we refer, as stated, to Cesari and Kannan [9, 10]. Parts of Sections 1 - 6 of this paper have been read at the May 1978 Conference in Florence, Italy, and distributed to the audience there in provisional form. Section 7 is completely new.

REFERENCES

[1] Bachman, G., and L. Narici, "Functional Analysis," Academic Press (1966).

[2] Cesari, L., *Functional analysis, nonlinear differential equations, and the alternative method.* "Nonlinear Analysis and Differential Equations," (L. Cesari, R. Kannan, J. D. Schuur, eds.) M. Dekker, New York, (1976), pp. 1-197.

[3] Cesari, L., *An abstract existence theorem across a point of resonance,* Dynamical Systems, an International Symposium at the Univ. of Florida (A. R. Bednarek and L. Cesari, eds.), Academic Press, (1977), pp. 11-26.

[4] Cesari, L., *Nonlinear oscillations across a point of resonance for nonselfadjoint systems*. Journ. Differential Equations, 28(1978), pp. 43-59.

[5] Cesari, L., *Nonlinear Analysis, a point of resonance for nonselfadjoint systems*, "Nonlinear Analysis," a volume in honor of E. H. Rothe, Academic Press (1978), pp. 43-67.

[6] Cesari, L., *Existence in the large of periodic solutions of hyperbolic partial differential equations*, Archive Rat. Mech. Anal., 20(1965), pp. 170-190.

[7] Cesari, L., and R. Kanna, *Functional analysis and nonlinear differential equations*, Bulletin Amer. Math. Soc. 63(1977), pp. 221-225.

[8] Cesari, L., and R. Kannan, *An abstract existence theorem at resonance*, Proc. Amer. Math. Soc., 63(1977), pp. 221-225.

[9] Cesari, L., and R. Kannan, *Existence of solutions of nonlinear hyperbolic problems*, Annali Scuola Normale Sup. Pisa. (4) 6 (1979), pp. 573-592.

[10] Cesari, L., and R. Kannan, *Solutions of nonlinear hyperbolic equations at resonance*, Journ. Nonlinear Analysis, (to appear).

[11] Cesari, L. and P. J. McKenna, *Alternative problems and Grothendieck approximation properties*, Bulletin Inst. Math. Acad. Sinica, Taiwan, 6(1978), pp. 569-581.

[12] Hale, J. K., *Periodic solutions of a class of hyperbolic equations*, Archive Rat. Mech. Anal. 23(1967), pp. 380-398.

[13] Hall, W. S., *On the existence of periodic solutions for the equation* $D_{tt}u + (-1)^p D_x^{2p} u = f(.,.,u)$, Journ. Differential Equations 7(1970), pp. 509-526.

[14] Hall, W. S., *Periodic solutions of a class of weakly nonlinear evolution equations*, Archive Rat. Analysis 39(1970), pp. 294-322.

[15] Kannan, R., and P. J. McKenna, *An existence theorem by alternative method for semilinear abstract equations*, Bollettino Unione Mat. Italiana (5) 14A (1977), pp. 355-358.

[16] McKenna, P. J., *Nonselfadjoint semilinear equations at multiple resonance in the alternative method*, Journ. Differential Equation. 33(1979), pp. 275-293.

[17] Petzeltova, H., *Periodic solutions of the equation* $u_{tt} + u_{xxxx} = f(.,.,u,u_t)$, Czechoslovak. Math. Journ. 23(98), pp. 269-285.

[18] Rothe, E. H., *On the Cesari index and the Brouwer-Petryshin degree*, Dynamical Systems, an International Symposium at the University of Florida (A. R. Bednarek and L. Cesari eds.), Academic Press (1977), pp. 295-312.

STABILITY FROM THE BIFURCATION FUNCTION

Jack K. Hale[1]

Brown University

INTRODUCTION

In a recent paper, deOliveira and Hale [2] discussed the relationship between the bifurcation function obtained by the method of Liapunov-Schmidt and the flow on the center manifold for periodic n-dimensional systems for which the linear approximation has one characteristic multiplier equal to one and the remaining ones inside the unit circle. The purpose of this paper is to present a more elementary proof of this result and also to give an application to stability for autonomous systems in the critical case of one zero root. This allows one to extend and simplify the classical results of Liapunov [4] to the case of C^k-vector fields. The approach used by Liapunov cannot be extended to this case since he employs the theory of transformations to approximate the vector fields by simpler vector fields.

Consider the n-dimensional vector equation

$$\dot{x} = AX + F(t,x,\lambda), \quad A = \begin{pmatrix} 0 & 0 \\ 0 & B \end{pmatrix} \tag{1}$$

[1]This research was supported in part by the National Science Foundation under MCS-79-05774, in part by the United States Army under AROD DAAG 27-76-G-0294, and in part by the United States Air Force under AF-AFOSR 76-3092C.

ISBN 0-12-045550-1

where $\lambda \in E$, a Banach space, B is an $(n - 1) \times (n - 1)$ matrix with eigenvalues with negative real parts, $F(t,x,\lambda)$ is continuous, is C^k, $k \geq 1$, in x,λ, $F(t + 1,x,\lambda) = F(t,x,\lambda)$, $F(t,0,0) = 0, \partial F(t,0,0)/\partial x = 0$. Our objective is to determine the 1-periodic solutions of (1) in a neighborhood of $(x,\lambda) = (0,0)$ as well as their stability properties.

The method of Liapunov-Schmidt is an effective way of determining the number of 1-periodic solutions of (1). Since A has 0 as a simple eigenvalue and all other eigenvalues with negative real parts, this method yields a scalar function $G(a,\lambda)$ defined and C^k for (a,λ) in a neighborhood of $(0,0) \in \mathbb{R} \times E$ such that Eq. (1) has a 1-periodic solution in a certain neighborhood of $(x,\lambda) = (0,0)$ if and only if $G(a,\lambda) = 0$. Also, there is a prescription for determining the 1-periodic solutions from the zeros of G. The method of Liapunov-Schmidt can be considered as a reduction principle - reducing the discussion of the existence of special solutions of (1) (the 1-periodic solutions) to an alternative problem in lower dimension - namely, to finding the zeros of the scalar function $G(a,\lambda)$.

There is another reduction principle in differential equations which determines the behavior of all solutions of (1) in a neighborhood of $(x,\lambda) = 0$. This is the center manifold theorem. If $x = (y,z)$, $y \in \mathbb{R}$, $z \in \mathbb{R}^{n-1}$ in Eq. (1), this theorem ways there is a function $h(t,y,\lambda) \in \mathbb{R}^{n-1}$, defined, continuous, C^k in y,λ in $\mathbb{R} \times U$, where U is a neighborhood of $(y,\lambda) = (0,0) \in \mathbb{R} \times E$, such that $h(t,y,\lambda) = h(t + 1,y,\lambda)$ and the set

$$M_\lambda \overset{\text{def}}{=} \{(t,y,z) : y \in \mathbb{R},\ z \in \mathbb{R}^{n-1},\ z = h(t,y,\lambda)\ t \in \mathbb{R},\ (y,\lambda) \in U\}$$

is an invariant manifold for Eq. (1). Furthermore, this manifold M_λ is exponentially asymptotically stable. The properties of the solutions of Eq.(1) in a neighborhood of $(x,\lambda) = (0,0)$ is completely determined by the flow induced on M_λ by the scalar equation

$$\dot{y} = f(t,y,h(t,y,\lambda),\lambda) \tag{2}$$

where $F = (f,g)$, $f \in \mathbb{R}$, $g \in \mathbb{R}^{n-1}$. Any 1-periodic solution of Eq. (1) near $(x,\lambda) = (0,0)$ is given by $x(t) = (y(t), z(t))$, $z(t) = h(t,y(t),\lambda)$, and $y(t)$ a 1-periodic solution of (2), and conversely. Furthermore, the stability properties of $y(t)$ determine the stability properties of $x(t)$ is the sense that they are the same except that $x(t)$ always has an additional $(n - 1)$-dimensional stable manifold.

The principle result proved by deOliveira and Hale [2] is

Theorem 1. Consider the scalar equation

$$\dot{a} = G(a,\lambda) \tag{3}$$

where $G(a,\lambda)$ is the bifurcation function obtained by the method of Liapunov-Schmidt. The zeros of $G(a,\lambda)$ are in one-to-one correspondence with the 1-periodic solutions of Eq. (2) in a neighborhood of $(y,\lambda) = (0,0)$. Furthermore, the stability properties of the equilibrium points of (3) are the same as the stability properties of the 1-periodic solutions of (2).

Proof: Let us indicate a proof which is different from and simpler than the one in [2]. We first prove it for the scalar equation (2). For Eq. (2), one can also apply the method of Liapunov-Schmidt to obtain a scalar bifurcation $\tilde{G}(a,\lambda)$ in the following way. In a neighborhood of $(y,a,\lambda) = (0,0,0)$, let $y(t,a,\lambda)$ be the unique 1-periodic solution of the equation

$$\dot{y}(t) = f(t,y(t),h(t,y(t),\lambda)) - \int_0^1 f(s,y(s),h(\dot{s},y(s),\lambda),\lambda)ds \tag{4}$$

with $\int_0^1 y = a$ and define

$$\tilde{G}(a,\lambda) = \int_0^1 f(s,y(s,a,\lambda),h(s,y(s,a,\lambda),\lambda)ds. \tag{5}$$

The 1-periodic solutions of Eq. (2) in a neighborhood of zero are in one-to-one correspondence with the zeros of $\tilde{G}$ in a neighborhood of zero. Let us make the transformation of variables

$$y \to b : y(t) = y(t,b,\lambda)$$

in Eq. (2). Since $y(t,b,0) = b$, $\partial y(t,0,0)/\partial b = 1$, and $y(t,b,\lambda)$ satisfies (4), (5), we have

$$\dot{b} = (\partial y(t,b,\lambda)/\partial b)^{-1}\tilde{G}(b,\lambda). \tag{6}$$

For (b,λ) in a sufficiently small neighborhood of zero, $(\partial y(t,b,\lambda)/\partial b)^{-1} > 0$. This immediately shows that the stability properties of any 1-periodic solution of (6) (which coincides with the zeros of $\tilde{G}$) is the same as the stability properties of the corresponding solution of $\dot{a} = \tilde{G}(a,\lambda)$. This proves the theorem for Eq. (2).

To prove the theorem for Eq. (1), we proceed as in [2], first observing that $\tilde{G}(a,\lambda), G(a,\lambda)$ have the same set of zeros in a neighborhood of zero. Thus, we need only show that $\tilde{G}, G$ have the same sign between zeros. Suppose this is not the case. By a small perturbation in the original

Eq. (1), we can suppose the zeros of both functions are simple. Now replace f in Eq. (1) by $f + \varepsilon$, obtain the new bifurcation $\tilde{G}(a,\lambda,\varepsilon) = G(a,\lambda) + \varepsilon$ for (1), obtain the new bifurcation function $\tilde{G}(a,\lambda,\varepsilon) = \tilde{G}(a,\lambda) + \delta(a,\lambda)\varepsilon + 0(|\varepsilon|^2)$ as $|\varepsilon| \to 0$, $\delta(a,\lambda) > 0$, for the corresponding equation on the center manifold. Now suppose (a_0,λ_0) is a simple zero of $G(a,\lambda)$, $\tilde{G}(a,\lambda)$ with $(\partial G/\partial a)(\delta\tilde{G}/\delta a) < 0$ at (a_0,λ_0); that is, the functions have opposite sign in a neighborhood of (a_0,λ_0). The functions $G(a,\lambda,\varepsilon)$,$\tilde{G}(a,\lambda,\varepsilon)$ have distinct zeros in a neighborhood of (a_0,λ_0) for ε small. But this is a contradiction since they must have the same zeros. This proves the theorem.

Applications of Theorem 1 to general bifurcation theory and to Hopf bifurcation may be found in [2]. Also, the extension to infinite dimensional systems is discussed - especially parabolic equations and functional differential equations.

Let us give another application to the classical problem of stability in critical cases discussed so thoroughly by Liapunov [4], Lefschetz [3] and, more recently, by Bibikov [1] as well as others.

Consider the equation

$$\begin{aligned} \dot{y} &= f(y,z) \\ \dot{z} &= Bz + g(y,z) \end{aligned} \tag{7}$$

with $y \in \mathbb{R}$, $z \in \mathbb{R}^{n-1}$, B an $(n - 1) \times (n - 1)$ matrix whose eigenvalues have negative real parts, f,g, are C^k functions vanishing together with their first derivatives at $(0,0)$.

The method of Liapunov-Schmidt applied to this equation yields the bifurcation function

$$G(a) = f(a,\phi(a)) \tag{8}$$

where ϕ is the unique solution of the equation

$$B\phi + g(a,\phi) = 0 \tag{9}$$

in a neighborhood of $(a,\phi) = (0,0)$.

As immediate consequences of Theorem 1, we have

Corollary 1. If there is an integer $q \geq 2$ and $\beta \neq 0$ such that

$$G(a) = \beta a^q + O(|a|^{q+1}) \quad \text{as} \quad |a| \to 0,$$

then the solution $x = (y,z)$ of (7) is asymptotically stable if and only if $\beta < 0$, q odd. Otherwise, it is unstable.

This is the classical result of Liapunov. The function $G(a)$ can be C^∞ and there may never exist a β, q as in Corollary 1. Theorem 1 actually implies a more general property.

Corollary 2. The solution $(y,z) = (0,0)$ of (7) is asymptotically stable if and only if there is an $\varepsilon > 0$ such that $G(a)a < 0$ for $0 < |a| < \varepsilon$. The solution $(y,z) = (0,0)$ is unstable if and only if there is an $\varepsilon > 0$ such that $G(a)a > 0$ for either $0 < a < \varepsilon$ or $-\varepsilon < a < 0$. If there is an $\varepsilon > 0$ such that $G(a) = 0$ for $0 \leq |a| < \varepsilon$, then the solution $(y,z) = (0,0)$ of (7) is stable and there is a first integral in a neighborhood of $(0,0)$.

Proof: Everything is obvious from Theorem 1 except the existence of the first integral. Suppose $G(a) = 0$ for $|a| < \varepsilon$ and let $z = h(y)$ be the parametric representation of a center manifold M at $(0,0) : M = \{(y,z) : z = h(y)\}$. We know that $f(y,h(y)) = 0$ for $|y| < \varepsilon$. Also, each equilibrium point $(y_0,h(y_0))$ of (7) $|y| < \varepsilon$, has an (n - 1)-dimensional stable manifold $S(y_0)$ which is C^k in y_0. The curve M and these stable manifolds can be used as a coordinate system in a neighborhood of (0,0) to obtain a C^k mapping $H:(y,z) \to (u,v)$, where $\dot{u} = 0$, $\dot{v} = Bv + \tilde{g}(u,v)$. This latter equation has a first integral $V(u,v) = u$ so that the original equation has a first integral. This proves the result.

<u>Corollary 3</u>. If there is scalar function $H(y,z,w)$, continuous for $(y,z,w) \in \mathbb{R} \times \mathbb{R}^{n-1} \times \mathbb{R}^{n-1}$ such that $H(y,z,0) = 0$ and

$$f(y,z) = H(y,z,Bz + g(y,z))$$

then the zero solution of (1) is stable and there is a first integral.

<u>Proof</u>: The hypotheses imply $G(y) = f(y,\phi(y)) = 0$ for y in a neighborhood of zero and so Corollary 2 applies.

In the case of analytic systems, Bibikov [1] refers to the situation in Corollary 1 as the <u>algebraic case</u>. For f,g analytic, the function G(a) is analytic and therefore either the algebraic case holds or $G(a) \equiv 0$, which is called the <u>transcendental case</u>. Corollary 2 says there is a first integral in the analytic case - another classical result of Liapunov (see Bibikov]1]). Corollary 3 was also stated by Liapunov for analytic systems. Thus, we see that the basic results of Liapunov can be generalized to C^k-vector fields and, in addition, everything is based only on the bifurcation function. This latter remark is the essential improvement in the statement of the results of Liapunov. Some aspects of the proofs, however, are similar. Liapunov used his general transformation theory to put the equation in a form where it is easy to discover the center manifold and the flow on the center manifold. We use the abstract center manifold theory and properties of stable manifold. In addition, a small amount of perturbation theory is used in an abstract way in order to prove that the bifurcation function determines the stability properties of the solutions.

It is instructive for the reader to check the original examples given by Liapunov [4] to see how only the bifurcation function was used to determine stability.

We remark that the same results as above have extensions to certain evolutionary equations in infinite dimensions; for example, parabolic systems and functional differential equations.

Finally, we have emphasized stability of the solution $(y,z) = 0$ of (7). If the autonomous system depends on a parameter (which often occurs in applications), Theorem 1 may be applied directly to obtain stability results even at the bifurcation curves.

REFERENCES

[1] Bibikov, Yu.N., "Local Theory of Nonlinear Analytic Ordinary Differential Equation." Lecture Notes in Math, Vol. 702, Springer-Verlag, 1979.

[2] deOliveira, J. C., and J. K. Hale, *Dynamic bifurcation from bifurcation equations.* Tôhoku Math. J. To appear.

[3] Lefschetz, S., "Differential Equations - Geometric Theory," Wiley Interscience, 1963.

[4] Liapunov, A. M., "Problème Général de la Stabilité du Mouvement." Princeton, 1949.

BOUNDARY VALUE PROBLEMS FOR LIPSCHITZ EQUATIONS

Lloyd K. Jackson

University of Nebraska

INTRODUCTION

We shall be concerned with boundary value problems for a scalar equation

$$x^{(n)} = f(t,x,x',\ldots,x^{(n-1)}) \tag{1}$$

in which f is continuous on a slab $(a,b) \times R^n$ and satisfies a Lipschitz condition

$$|f(t,y_1,\ldots,y_n) - f(t,z_1,\ldots,z_n)| \le \sum_{j=1}^{n} k_j |y_j - z_j| \tag{2}$$

on the slab. As in [1] the discussion could be carried out in terms of more general Lipschitz conditions but for simplicity we shall confine ourselves to equations (1) which satisfy a Lipschitz condition of the form (2).

If particular boundary conditions are imposed, a standard problem is to determine an $h > 0$ such that the corresponding boundary value problem for any equation (1) satisfying (2) will have a unique solution on any $[c,d] \subset (a,b)$ with $d - c < h$. A very common procedure used in attacking such problems involves applying the Contraction Mapping Principle, for example, [2], [3], [4]. With this method one arrives at the conclusion that the boundary value problem has a unique solution on any interval $[c,d] \subset (a,b)$ with $d - c < h_0$

ISBN 0-12-045550-1

where h_0 is the positive root of an equation

$$\sum_{j=1}^{n} \lambda_j k_j h^{n+1-j} = 1$$

and the λ_j, $1 \le j \le n$, are constants associated with upper bounds of integrals of a suitable Green's function and its derivatives. The interval length h_0 obtained in this way has the advantage that on intervals of length less than h_0 the solution can be calculated by successive approximations. However, this value is not in general best possible in the sense that unique solutions may exist on longer intervals.

Recently, Melentsova and Milshtein [5] used Control Theory methods to find best possible interval lengths for existence and uniqueness of solutions of certain types of boundary value problems for linear differential equations with bounded coefficients. Subsequently, Melentsova [6] obtained similar results for linear equations with more general constraints on the coefficients. In [1] the results in [5] were generalized to higher order equations and applications were made to nonlinear equations.

In the present paper we shall correct some errors in [1] and clear up some questions that were left open. Also we shall consider some other types of boundary value problems than those dealt with in [1]. In Section 2 we shall specify the type of boundary value problem with which we will be dealing and observe how the Pontryagin Maximum Principle can be applied. In Section 3 we shall consider some questions of constancy of sign which arise in applying the Maximum Principle. In Section 4 we shall arrive at best possible interval lengths for existence and uniqueness of solutions of the boundary value problems under consideration, and apply these results to nonlinear equations.

2. The Maximum Principle. Let U be the set of all vector functions $u(t) = (u_1(t), \ldots, u_n(t)$ such that the component functions $u_j(t)$ are Lebesgue measurable on (a,b) and satisfy the inequalities

$$|u_j(t)| \le k_j$$

on (a,b) for $1 \le j \le n$. Let I and J be nonnull subsets of the set of integers $\{1,2,\ldots,n\}$ such that, if there are k elements in I, then there are $n - k$ elements in J. We do not assume $I \cap J = \phi$. Let I^c and J^c be the respective complements of I and J in $\{1,2,\ldots,n\}$. For fixed such sets of integers I and J we consider the boundary value problems

$$x^{(n)} = \sum_{j=1}^{n} u_j(t) x^{(j-1)} \tag{3}$$

$$x^{(i-1)}(c) = 0 \quad \text{for} \quad i \in I, \text{ and} \tag{4}$$

$$x^{(i-1)}(d) = 0 \quad \text{for} \quad i \in J, \tag{5}$$

where $a < c < d < b$ and $u = (u_1(t),\ldots,u_n(t)) \in U$. If there exist $a < c < d < b$ and $u \in U$ such that the corresponding problem (3), (4), (5) has a nontrivial solution, then it follows from standard arguments that a time optimal solution exists. That is, there is a $u^* \in U$ and $c \le c_1 < d_1 \le d$ such that

$$x^{(n)} = \sum_{j=1}^{n} u_j^*(t) x^{(j-1)}$$

has a nontrivial solution $x(t)$ with

$$x^{(i-1)}(c_1) = 0 \quad \text{for} \quad i \in I,$$

$$x^{(i-1)}(d_1) = 0 \quad \text{for} \quad i \in J,$$

and $d_1 - c_1$ is a minimum over all such solutions. For this time optimal solution $x(t)$ let $z(t) = (x(t), x'(t), \ldots, x^{(n-1)}(t))^T$. Then $z(t)$ is a solution of the first order system

$$z' = A[u^*(t)]z$$

corresponding to the scalar equation

$$x^{(n)} = \sum_{j=1}^{n} u_j^*(t)x^{(j-1)}.$$

The Pontryagin Maximum Principle [7,p.314] asserts that there is a nontrivial solution $\psi(t) = (\psi_1(t), \ldots, \psi_n(t))^T$ of the adjoint system

$$\psi' = -A^T[u^*(t)]\psi \tag{6}$$

such that

$$\begin{aligned} \sum_{j=1}^{n} x^{(j)}(t)\psi_j(t) &= (z'(t), \psi(t)) \\ &= \operatorname{Max}\{A[u(t)]z(t), \psi(t)) : u \in U\} \end{aligned} \tag{7}$$

for almost all t with $c_1 \le t \le d_1$, where $(.,.)$ is the inner product. Furthermore, $(z'(t), \psi(t))$ is a nonnegative constant for almost all $c_1 \le t \le d_1$,

$$\psi_j(c_1) = 0 \quad \text{for} \quad j \in I^c,$$

and

$$\psi_j(d_1) = 0 \quad \text{for} \quad j \in J^c.$$

Since

$$(A[u(t)]z(t)\psi(t)) = \sum_{j=1}^{n-1} x^{(j)}(t)\psi_j(t)$$

$$+ \psi_n(t) \sum_{j=1}^{n} u_j(t)x^{(j-1)}(t),$$

the maximum condition (7) can be written as

$$\psi_n(t) \sum_{j=1}^{n} u_j^*(t)x^{(j-1)}(t) = \tag{8}$$

$$\text{Max}\{\psi_n(t) \sum_{j=1}^{n} u_j(t)x^{(j-1)}(t) : u \in U\}.$$

If $\psi_n(t)$ changes sign on (c_1,d_1), the condition (8) appears to be of little use. However, if $\psi_n(t)$ has no zeros on (c_1,d_1) and if $x(t) > 0$ on (c_1,d_1), then (8) determines the optimal control $u^*(t)$. In particular, if $x(t) > 0$ and $\psi_n(t) < 0$ on (c_1,d_1), then

$$u_1^*(t) = -k_1, \tag{9}$$

and for $2 \le j \le n$

$$u_j^*(t) = \begin{cases} -k_j & \text{when } x^{(j-1)}(t) \ge 0 \\ +k_j & \text{when } x^{(j-1)}(t) < 0. \end{cases} \tag{10}$$

on the otherhand, if $x(t) > 0$ and $\psi_n(t) > 0$ on (c_1,d_1), then

$$u_j^*(t) = +k_1, \tag{11}$$

and for $2 \le j \le n$

$$u_j^*(t) = \begin{cases} +k_j & \text{when } x^{(j-1)}(t) \ge 0 \\ -k_j & \text{when } x^{(j-1)}(t) < 0. \end{cases} \tag{12}$$

It follows that, if $x(t) > 0$ and $\psi_n(t) < 0$ on (c_1,d_1), then the time optimal solution $x(t)$ is a solution of

$$x^{(n)} = -k_1 x - \sum_{j=2}^{n} k_j |x^{(j-1)}| \tag{13}$$

on $[c_1, d_1]$, and, if $x(t) > 0$ and $\psi_n(t) > 0$ on (c_1, d_1), then the time optimal solution $x(t)$ is a solution of

$$x^{(n)} = k_1 x + \sum_{j=2}^{n} k_j |x^{(j-1)}(t)| \tag{14}$$

on $[c_1, d_1]$.

If there is a $u(t) \in U$ such that the problem (3), (4), (5) has a nontrivial solution, then the problem

$$\psi' = -A^T[u(t)]\psi \tag{15}$$

$$\psi_j(c) = 0 \quad \text{for} \quad j \in I^c \tag{16}$$

$$\psi_j(d) = 0 \quad \text{for} \quad j \in J^c \tag{17}$$

also has a nontrivial solution, and conversely. Thus, the Maximum Principle associates with a time optimal solution of (3), (4), (5) a time optimal solution of (15), (16), (17), and conversely.

3. Constancy of Sign for Time Optimal Solutions. In the preceeding section we noted that, if the time optimal solution of a boundary value problem (3), (4), (5) and the associated solution of the adjoint system satisfy certain sign conditions, then the time optimal solution is a solution of either (13) or (14). Our method of establishing that the desired sign conditions hold requires that we look at successive problems in certain collections of boundary value problems. In particular, we shall consider boundary value problems (3), (4), (5) in which boundary conditions (4) and (5) are

$$x^{(i-1)}(c) = 0 \quad \text{for} \quad 1 \le i \le k$$

and $x^{(i-1)}(d) = 0$ for $1 \le i \le n - k$,

where $1 \le k \le n - 1$. We shall refer to these boundary conditions as $(k,n - k)$ conjugate boundary conditions. If an equation (3) has no nontrivial solution satisfying these boundary conditions for any $c < d$ in an interval, we will say that the equation is $(k,n - k)$ disconjugate on that interval. We shall also consider boundary value problems (3), (4), (5) in which the boundary conditions (4), (5) are

$$x^{(i-1)}(c) = 0 \quad \text{for } 1 \le i \le k$$

and $x^{(i-1)}(d) = 0$ for $k + 1 \le i \le n$,

where $1 \le k \le n - 1$. These boundary conditions will be called $(k,n - k)$ focal boundary conditions and an equation (3) will be called $(k,n - k)$ disfocal on an interval if it has no nontrivial solutions satisfying these conditions on that interval.

Theorem 1. If there is a control vector $u(t) \in U$ such that the corresponding equation (3) has a nontrivial solution satisfying $(n - 1,1)$ conjugate boundary conditions on (a,b) and if $x(t)$ is a time optimal solution with

$$x^{(i-1)}(c) = 0 \text{ for } 1 \le i \le n - 1,$$
$$x(d) = 0,$$

and with $d - c$ a minimum, then $x(t)$ is a solution of (13) on $[c,d]$. If for all $u(t) \in U$ and all j with $k + 1 \le j \le n - 1$ the corresponding equations (3) are $(j,n - j)$ disconjugate on (a,b) and if there is an equation in the collection (3) which has a nontrivial solution satisfying $(k,n - k)$ conjugate boundary conditions on (a,b), then a time optimal solution $x(t)$ with

$$x^{(i-1)}(c) = 0 \quad \text{for} \quad 1 \le i \le k,$$

$$x^{(i-1)}(d) = 0 \quad \text{for} \quad 1 \le i \le n-k,$$

and $d - c$ a minimum is a solution of (13) on $[c,d]$ when $n - k$ is odd and is a solution of (14) on $[c,d]$ when $n - k$ is even.

Proof. First assume that $x(t)$ is a time optimal solution satisfying the $(n - 1, 1)$ conjugate boundary conditions

$$x^{(i-1)}(c) = 0 \quad \text{for} \quad 1 \le i \le n-1$$

and $$x(d) = 0$$

with $d - c$ a minimum. Then because of the time optimality $x(t) \ne 0$ for $c < t < d$ and without loss of generality we can assume $x(t) > 0$ on (c,d). It follows that $x^{(n-1)}(c) > 0$. Since the solution $\psi(t)$ of the adjoint system associated with $x(t)$ by the Maximum Principle satisfies

$$\psi_n(c) = 0$$

and $\psi_i(d) = 0$ for $2 \le i \le n$,

and is also time optimal, it follows that $\psi_n(t) \ne 0$ for $c < t < d$. Thus $x(t)$ is a solution of either (13) or (14) on $[c,d]$ which means that in either case $x^{(n-1)}(t)$ is strictly mònotone on $[c,d]$. Now we also have that

$$\sum_{j=1}^{n} x^{(j)}(t)\psi_j(t) = x^{(n-1)}(c)\psi_{n-1}(c) = x'(d)\psi_1(d)$$

on $c \le t \le d$ with the constant value being nonnegative. If $x'(d) = 0$, then repeated applications of Rolle's Theorem produces two zeros of $x^{(n-1)}(t)$ on (c,d) which contradicts the strict monotoneity of $x^{(n-1)}(t)$. Also, $\psi_1(d) \ne 0$

since $\psi(t)$ is a nontrivial solution of the adjoint system. Thus, $x'(d)\psi_1(d) \neq 0$ and $x^{(n-1)}(c)\psi_{n-1}(c) > 0$. Since $x^{(n-1)}(c) > 0, \psi_{n-1}(c) > 0$ and an examination of the adjoint system shows that this implies that $\psi_n(t) < 0$ on (c,d). Thus, $x(t)$ is a solution of (13) on $[c,d]$.

Now assume that for each $u(t) \in U$ the corresponding equation (3) is $(j,n-j)$ disconjugate on (a,b) for each j with $k+1 \leq j \leq n-1$; but assume that there is an equation (3) with a nontrivial solution satisfying $(k,n-k)$ conjugate boundary conditions. Let $x(t)$ be a time optimal such solution with

$$x^{(i-1)}(c) = 0 \quad \text{for} \quad 1 \leq i \leq k,$$

$$x^{(i-1)}(d) = 0 \quad \text{for} \quad 1 \leq i \leq n-k,$$

and $d - c$ a minimum.

Then it follows from Lemma 4 in [8] that $x(t) \neq 0$ for $c < t < d$ and again without loss of generality we can assume $x(t) > 0$ on (c,d). Since all equations (3) with $u(t) \in U$ are $(k+1,n-k-1)$ disconjugate on (a,b), it follows that $x^{(k)}(c) \neq 0$. Thus, $x^{(k)}(c) > 0$. In the Corollary to Theorem 6 in [1] it is proven that $\psi_n(t) \neq 0$ in (c,d) where $\psi(t)$ is the solution of the adjoint system associated with $x(t)$ by the Maximum Principle. Hence, again $x(t)$ is a solution of either (13) or (14) on $[c,d]$. As in the proof of the first part of the Theorem, this implies $x^{(n-k)}(d) \neq 0$. Furthermore, because of the adjoint system having only the trivial solution satisfying

$$\psi_i(c) = 0 \quad \text{for} \quad k+2 \leq i \leq n$$

and $\psi_i(d) = 0$ for $n-k \leq i \leq n$,

we conclude that $\psi_{n-k}(d) \neq 0$ for the solution of the adjoint system associated with $x(t)$. Thus

$$\sum_{j=1}^{n} x^{(j)}(t)\psi_j(t) = x^{(k)}(c)\psi_k(c) = x^{(n-k)}(d)\psi_{n-k}(d) \neq 0$$

and from $x^{(k)}(c) > 0$ we conclude that $\psi_k(c) > 0$. An examination of the adjoint system leads to the conclusion that $\psi_j(c) = 0$ for $k + 1 \leq j \leq n$ and $\psi_k(c) > 0$ implies sign $\psi_n(t) = (-1)^{n-k}$ on (c,d). Thus $x(t)$ is a solution of (13) on $[c,d]$ if $n - k$ is odd and is a solution of (14) on $[c,d]$ if $n - k$ is even.

Theorem 2. if there is a control $u(t) \in U$ such that the associated equation (3) has a nontrivial solution satisfying $(n - 1,1)$ focal boundary conditions on (a,b) and if $x(t)$ is a time optimal solution with

$$x^{(i-1)}(c) = 0 \quad \text{for} \quad 1 \leq i \leq n - 1$$

$$x^{(n-1)}(d) = 0,$$

and $d - c$ a minimum, then $x(t)$ is a solution of (13) on $[c,d]$. If for all $u(t) \in U$ and all j with $k + 1 \leq j \leq n - 1$ the corresponding equations (3) are $(j,n - j)$ disfocal on (a,b) and if there is an equation in the collection (3) which has a nontrivial solution satisfying $(k,n - k)$ focal boundary conditions on (a,b), then a time optimal solution $x(t)$ with

$$x^{(i-1)}(c) = 0 \quad \text{for} \quad 1 \leq i \leq k,$$

$$x^{(i-1)}(d) = 0 \quad \text{for} \quad k + 1 \leq i \leq n,$$

and $d - c$ a minimum is a solution of (13) on $[c,d]$ if $n - k$ is odd and is a solution of (14) on $[c,d]$ if $n - k$ is even.

Proof. Let $x(t)$ be a time optimal solution with

$$x^{(i-1)}(c) = 0 \quad \text{for} \quad 1 \le i \le n-1$$

$$x^{(n-1)}(d) = 0,$$

and $d - c$ a minimum. Then the associated solution $\psi(t)$ of the adjoint system is a time optimal solution satisfying the conditions

$$\psi_n(c) = 0,$$

and $\psi_j(d) = 0$ for $1 \le j \le n-1$.

It follows that $\psi_n(t) \ne 0$ on $(c,d]$. Since $x^{(n-1)}(c) \ne 0$ we can assume $x^{(n-1)}(c) > 0$ and $x(t) > 0$ on $(c,d]$. Thus $x(t)$ is a solution of either (13) or (14) on $[c,d]$ and it is clear that it must be a solution of (13).

Now assume that for each $u(t) \in U$ the corresponding equation (3) is $(j, n-j)$ disfocal on (a,b) for each $k+1 \le j \le n-1$ but that there is an equation (3) with a non-trivial solution satisfying $(k, n-k)$ focal boundary conditions on (a,b). Let $x(t)$ be a time optimal such solution with

$$x^{(i-1)}(c) = 0 \quad \text{for} \quad 1 \le i \le k,$$

$$x^{(i-1)}(d) = 0 \quad \text{for} \quad k+1 \le i \le n,$$

and $d - c$ a minimum. For the equation (3) of which $x(t)$ is a solution let d_0 be the infimum of all s with $c < s < b$ such that there is a nontrivial solution $y(t)$ satisfying

$$y^{(i-1)}(c) = 0 \quad \text{for} \quad 1 \le i \le k$$

$$y^{(i-1)}(t_i) = 0 \quad \text{for} \quad k+1 \le i \le n,$$

where $c \le t_{k+1} \le \dots \le t_n \le s$. Then $c < d_0 \le d$. Using the same type of argument as was used by Muldowney in the proof of Proposition 1 in [9], one can prove that there is a nontrivial solution $y(t)$ such that for some m with $k \le m \le n - 1$

$$y^{(i-1)}(c) = 0 \quad \text{for} \quad 1 \le i \le m,$$

$$y^{(i-1)}(d_0) = 0 \quad \text{for} \quad m + 1 \le i \le n,$$

and $y^{(i-1)}(t) \ne 0$ on (c,d_0) for $1 \le i \le m$.
It follows from the $(j,n-j)$ disfocality for $k + 1 \le j \le n - 1$ that $m = k$. It then follows from the optimality of the solution $x(t)$ that $d_0 = d$ and then from the $(k + 1,n - k - 1)$ disfocality that the above extremal solution $y(t)$ is a scalar multiple of $x(t)$. Hence, we conclude that $x^{(i-1)}(t) \ne 0$ on (c,d) for $1 \le i \le k$. It follows that $x(t) \ne 0$ on $(c,d]$ since otherwise the preceeding assertion would be contradicted by repeated applications of Rolle's Theorem. We can assume then that $x^{(k)}(c) > 0$ and $x(t) > 0$ on $(c,d]$.

Now let $\psi(t)$ be the time optimal solution of the adjoint system associated with $x(t)$ by the Maximum Principle. Then

$$\psi_i(c) = 0 \quad \text{for} \quad k + 1 \le i \le n$$

and $\psi_i(d) = 0$ for $1 \le i \le k$.
If we reverse the order of the components of $\psi(t)$, that is, define the vector function $y(t) = (y_1(t), \dots, y_n(t))$ by $y_j(t) = \psi_{n+1-j}(t)$ for $1 \le j \le n$, then $y(t)$ is a solution of a first order system of the type considered by Hinton in [10]. Furthermore,

$$y_i(c) = 0 \quad \text{for} \quad 1 \le i \le n - k$$

and $y_i(d) = 0$ for $n - k + 1 \le i \le n$

so that

$$y(t) = \sum_{j=1}^{n-k} c_j y^j(t)$$

where $y^j(t)$ is the solution of our modified first order system such that

$$y_i^j(d) = \delta_{ij}, \quad 1 \le i,j \le n.$$

If in this representation of $y(t)$ the coefficient $c_{n-k} = 0$, then $y_i(d) = 0$ for $i = n - k$ and we conclude that $\psi(t)$ is a nontrivial solution of the adjoint system with

$$\psi_i(c) = 0 \quad \text{for} \quad k + 2 \le i \le n$$

and $\psi_i(d) = 0$ for $1 \le i \le k + 1$.

This in turn would imply the existence of a nontrivial solution $z(t)$ of our time optimal equation from the collection (3) with

$$z^{(i-1)}(c) = 0 \quad \text{for} \quad 1 \le i \le k + 1$$

and $z^{(i-1)}(d) = 0$ for $k + 2 \le i \le n$.

This would contradict the $(k + 1, n - k - 1)$ disfocality, hence, we conclude $c_{n-k} \ne 0$.

If $W(y^1, \ldots, y^j)(t)$ is the determinant in which the ith row is $(y_i^1(t), \ldots, y_i^j(t))$ for $1 \le i \le j$, then it follows that $W(y^1, \ldots, y^j)(t) \ne 0$ on (a,d) for $1 \le j \le n - k - 1$ because of the $(j, n - j)$ disfocality on (a,b) for $k + 1 \le j \le n - 1$. With these conditions satisfied, if

$y_1(t) = 0$ at some point in $(c,d]$, we can apply Theorem 2.1 of [10] successively as was done in Theorem 6 in [1] to reach the conclusion that $W(y^1,\ldots,y^{n-k})(t_0) = 0$ for some $c < t_0 < d$. This contradicts the time optimality of $\psi(t)$ and we conclude that $y_1(t) = \psi_n(t) \neq 0$ on $(c,d]$. Thus, the time optimal solution $x(t)$ of our $(k,n-k)$ focal boundary value problem is either a solution of (13) on $[c,d]$ or a solution of (14) on $[c,d]$. In either case, from the form of equations (13) and (14) we see that $x^{(n)}(d) \neq 0$ since $x(d) \neq 0$. It follows that on $[c,d]$ we have

$$\sum_{j=1}^{n} x^{(j)}(t)\psi_j(t) \equiv x^{(k)}(c)\psi_k(c) = x^{(n)}(d)\psi_n(d) \neq 0.$$

Hence, $x^{(k)}(c)\psi_k(c) > 0$ and $\psi_k(c) > 0$ which again as in Theorem 1 yields sign $\psi_n(t) = (-1)^{n-k}$ on (c,d). This completes the proof of Theorem 2.

4. Existence and uniqueness of solutions. In this Section we apply the results of the preceding Section to obtain results concerning the uniqueness and existence of solutions of boundary value problems.

Theorem 3. Let $h = \mathrm{Min}\{h_k : 1 \le k \le [\frac{n}{2}]\}$, where $[\cdot]$ is the greatest integer function and h_k is the smallest positive number such that there is a solution $x(t)$ of the boundary value problem

$$x^{(n)} = (-1)^k[k_1 x + \sum_{j=2}^{n} k_j |x^{(j-1)}|],$$

$$x^{(i-1)}(0) = 0 \text{ for } 1 \le i \le n-k,$$

and $x^{(i-1)}(h_k) = 0$ for $1 \le i \le k$,

with $x(t) > 0$ on $(0,h_k)$ or $h_k = +\infty$ if no such solution exists. Let $f(t,x,x',\ldots x^{(n-1)})$ be continuous on $(a,b) \times R^n$ and satisfy the Lipschitz condition

$$(2) \quad |f(t,y_1,\ldots,y_n) - f(t,z_1,\ldots,z_n)| \le \sum_{j=1}^{n} k_j|y_j - z_j|$$

on $(a,b) \times R^n$. Then the boundary value problem

$$x^{(n)} = f(t,x,x',\ldots,x^{(n-1)})$$

$$x^{(i-1)}(t_j) = c_{ji},\ 1 \le i \le m_j,\ 1 \le j \le k,$$

where $2 \le k \le n$, $a < t_1 < t_2 < \ldots < t_n < b$, $m_j \ge 1$ for $1 \le j \le k$, and $\sum_{j=1}^{k} m_j = n$, has a unique solution for any assignment of real numbers c_{ji} if $t_k - t_1 < h$. Furthermore, this result is best possible for the class of all differential equations which satisfy the Lipschitz condition (2).

Proof. First we note that it follows from replacing t by $-t$ that

$$x(n) = (-1)^k[k_1x + \sum_{j=2}^{n} k_j|x^{(j-1)}|]$$

$$x^{(i-1)}(0) = 0 \quad \text{for} \quad 1 \le i \le n-k,$$

$$x^{(i-1)}(h_k) = 0 \quad \text{for} \quad 1 \le i \le k,$$

and $x(t) > 0$ on $(0,h_k)$

has a solution if and only if

$$x^{(n)} = (-1)^{n-k}[k_1x + \sum_{j=2}^{n} k_j|x^{(j-1)}|]$$

$$x^{(i-1)}(0) = 0 \quad \text{for} \quad 1 \le i \le k,$$

$$x^{(i-1)}(h_k) = 0 \quad \text{for} \quad 1 \le i \le n-k,$$

and $x(t) > 0$ on $(0,h_k)$

has a solution. Thus, if $h = \mathrm{Min}\{h_k : 1 \le k \le [\frac{n}{2}]\}$ and if $(c,d) \subset (a,b)$ with $d - c < h$, then for any $u(t) \in U$ the corresponding equation (3) is $(k, n - k)$ disconjugate on (c,d) for any k with $1 \le k \le n - 1$.

If $x(t)$ and $y(t)$ are distinct solutions of $x^{(n)} = f(t,x,x',\ldots,x^{(n-1)})$ such that $x(t_i) = y(t_i)$ for $1 \le i \le n$ where $a < t_1 < \ldots < t_n < b$, then as observed in Section 2 of [1] the difference $w(t) = x(t) - y(t)$ is a nontrivial solution of an equation (3) for a suitable $u^0(t) \in U$ and $w(t_i) = 0$ for $1 \le i \le n$. In this case Sherman [11] has proven that there is a subinterval $[c,d] \subset [t_1,t_n]$ and an integer k with $1 \le k \le n - 1$ such that the boundary value problem

$$x^{(n)} = \sum_{j=1}^{n} u_j^0(t) x^{(j-1)}$$

$$x^{(i-1)}(c) = 0 \quad \text{for} \quad 1 \le i \le k$$

$$x^{(i-1)}(d) = 0 \quad \text{for} \quad 1 \le i \le n - k$$

has a nontrivial solution. It was just observed that, if $t_n - t_1 < h$, this is impossible. Hence, on subintervals of (a,b) of length less than h solutions of n-point boundary value problems for $x^{(n)} = f(t,x,x',\ldots,x^{(n-1)})$, when they exist, are unique. Then, as pointed out in Section 1 of [1], on all subintervals of (a,b) of length less than h all n-point boundary value problems do have solutions which in turn implies that on subintervals of length less then h all boundary value problems of the type given in Theorem 3 have solutions which are unique.

For $1 \le k \le [\frac{n}{2}]$ let δ_k be the smallest positive number such that there is a solution $x(t)$ of the focal boundary value problem

$$x^{(n)} = (-1)^k[k_1 x + \sum_{j=2}^{n} k_j|x^{(j-1)}|],$$

$$x^{(i-1)}(0) = 0 \quad \text{for} \quad 1 \le i \le n-k,$$

and $x^{(i-1)}(\delta_k) = 0$ for $n-k+1 \le i \le n$,

with $x(t) > 0$ on $(0,\delta_k)$ or $\delta_k = +\infty$ if no such solution exists. For $[\frac{n}{2}] + 1 \le k \le n-1$ let δ_k be the smallest positive number such that there is a solution $x(t)$ of the problem

$$x^{(n)} = (-1)^{n+k}[k_1 x + \sum_{j=2}^{n} k_j|x^{(j-1)}|],$$

$$x^{(i-1)}(0) = 0 \quad \text{for} \quad n-k+1 \le i \le n,$$

and $x^{(i-1)}(\delta_k) = 0$ for $1 \le i \le n-k$,

with $x(t) > 0$ on $(0,\delta_k)$ or $\delta_k = +\infty$ if no such solution exists.

Theorem 4. Let $\delta = \text{Min}\{\delta_k : 1 \le k \le n-1\}$. Then, if $f(t,x,x',\ldots,x^{(n-1)})$ is continuous and satisfies the Lipschitz condition (2) on the slab $(a,b) \times R^n$, if $x(t)$ and $y(t)$ are solutions of $x^{(n)} = f(t,x,x',\ldots,x^{(n-1)})$ such that $x^{(i-1)}(t_i) = y^{(i-1)}(t_i)$ for $1 \le i \le n$ where $a < t_1 \le t_2 \le \ldots \le t_n < b$, and if $t_n - t_1 < \delta$, it follows that $x(t) \equiv y(t)$ on (a,b). Again, this result is best possible.

Proof. As in the proof of Theorem 3, if $x(t) \not\equiv y(t)$, then $w(t) = x(t) - y(t)$ is a nontrivial solution of an equation from the collection (3) with

$$w^{(i-1)}(t_i) = 0 \quad \text{for} \quad 1 \le i \le n.$$

It then follows from Proposition 1 of [9] that for that equation from the collection (3) there is a nontrivial solution $z(t)$, a $\bar{t}$ with $t_1 < \bar{t} \le t_n$, and an integer k with $1 \le k \le n$ such that

$$z^{(i-1)}(t_1) = 0 \quad \text{for} \quad 1 \le i \le k$$

and $z^{(i-1)}(\bar{t}) = 0$ for $k + 1 \le i \le n$.

However, from our choice of δ and from Theorem 2, it follows that this is impossible if $t_n - t_1 < \delta$.

In this case we do not have a "uniquness implies existence" theorem to appeal to since this is an open question for this type of focal boundary value problem. However, uniqueness does imply existence for linear differential equations so the following Corollary can be stated.

<u>Corollary 5</u>. Assume that $p_j(t)$, $1 \le j \le n$, and $q(t)$ are continuous on (a,b) and that

$$|p_j(t)| \le k_j \quad \text{on } (a,b)$$

for each $1 \le j \le n$. Then, if $t_n - t_1 < \delta$, the boundary value problem

$$x^{(n)} = \sum_{j=1}^{n} P_j(t)x^{(j-1)} + q(t)$$

$$x^{(i-1)}(t_i) = c_i, \quad 1 \le i \le n,$$

where $a < t_1 \le t_2 \le \dots \le t_n < b$ has a unique solution for any assignment of the boundary values c_i, $1 \le i \le n$.

REFERENCES

[1] Jackson, L., *Existence and uniqueness of solutions of boundary value problems for Lipschitz equations*, J. Differential Equations 32(1979), pp. 76-90.

[2] Bailey, P., L. Shampine, and P. Waltman, "Nonlinear Two Point Boundary Value Problems," Academic Press, New York, 1968.

[3] Barr, D., and T. Sherman, *Existence and uniqueness of solutions of three-point boundary value problems*, J. Differential equations 13(1973), pp. 197-212.

[4] Agarwal, R., and P. Krishnamurthy, *On the uniqueness of solutions of nonlinear boundary value problems*, J. Math. Phys. Sci. 10(1976), pp. 17-31.

[5] Melentsova, Y., and H. Milshtein, *An optimal estimate of the interval on which a multipoint boundary value problem has a solution*, Differencial'nye Uravrnenija 10(1974), pp. 1630-1641.

[6] Melentsova, Y., *A best possible estimate of the non-oscillation interval for a linear differential equation with coefficients bounded in* L_r, Differencial'nye Uravrnenija 13(1977), pp. 1776-1786.

[7] Lee, E., and L. Markus, "Foundations of Optimal Control Theory," Wiley, New York, 1967.

[8] Peterson, A., *Comparison theorems and existence theorems for ordinary differential equations*, J. Math. Anal. Appl. 55(1976), pp. 773-784.

[9] Muldowney, J., *A necessary and sufficient condition for disfocality*, Proc. Amer. Math. Soc. 74(1979), pp. 49-55.

[10] Hinton, D., *Disconjugate properties of a system of differential equations*, J. Differential Equations 2(1966), pp. 420-437.

[11] Sherman, T., *Properties of solutions of nth order linear equations*, Pacific J. Math. 15(1965), pp. 1045-1060.

PERIODIC SOLUTIONS OF SYSTEMS OF ORDINARY DIFFERENTIAL EQUATIONS*

S. J. Skar
R. K. Miller
A. N. Michel

Iowa State University

INTRODUCTION

The existence or nonexistence of oscillations is of fundamental importance in the design and the evaluation of feedback systems. Mees and Bergen [4] have recently proved interesting results on existence and nonexistence of periodic solutions of certain autonomous ordinary differential equations. The purpose of this paper is to extend some of their results to a class of interconnected systems. Following the general approach in Michel and Miller [5] we accomplish our analysis by viewing such systems as the interconnection of simpler subsystems. Each subsystem is analyzed using describing function techniques.

Our results address the most popular uses of describing functions - that is, we give conditions which insure that the system cannot sustain a π-symmetric oscillation. Our conditions are computable in the sense that certain parameters are obtained for each subsystem by graphical methods. These

*This research was supported by the National Science Foundation under grant ENG77-28446.

ISBN 0-12-045550-1

parameters are combined with parameters which measure the strength of the interconnections in order to form a test matrix. The test matrix must have positive successive principle minors in Theorem 1 and must satisfy certain diagonal dominance conditions in Theorem 2.

Background. Let R be the real line and R^{ℓ} be real ℓ-dimensional space with elements $x = (x_1, \ldots, x_{\ell})^T$ and norm $|x| = \max\{|x_k| : 1 \leq k \leq \ell\}$. The corresponding matrix norm is $\|A\| = \|\{a_{kj}\}\| = \max_k \sum_{j=1}^{\ell} |a_{kj}|$. We call an $\ell \times \ell$ matrix $A = \{a_{kj}\}$ an M-matrix if $a_{kj} \leq 0$ for $k \neq j$ and if the successive principle minors of A are all positive. It is known that if A is an M-matrix then $a_{kk} > 0$ for all $k = 1, 2, \ldots, \ell$ and all elements of A^{-1} are nonnegative, c.f. e.g. [1 or 5, Chapter 2].

Define $H(\omega)$ to be the set of all square integrable functions $\phi : [0, 2\pi/\omega] \to R$ which satisfy

(i) $\phi(t + 2\pi/\omega) = \phi(t)$ a.e. on R, and

(ii) $\phi(t + \pi/\omega) = -\phi(t)$ a.e. on R.

Property (i) states that ϕ is $2\pi/\omega$-periodic. Property (ii) is called π-symmetric in [4], odd-harmonic in [3] (since it implies that any nonzero Fourier coefficient of ϕ must be odd index), and ω/π-odd in [6] (since ϕ is odd about the point $t = \omega/\pi$). The definition $H(\omega)$ is easily extended to vector valued functions $\phi : R \to R^{\ell}$ for which each component satisfies (i) and (ii) above. For $\phi \in H(\omega)$ let

$$\phi(t) \sim \sum_{n \text{ odd}} \phi_n \exp(in\omega t)$$

be the Fourier series for ϕ. Then

$$\|\phi\|^2 = 2\sum_{n\ \mathrm{odd}}|\phi_n|^2 = \frac{\omega}{\pi}\int_0^{2\pi/\omega} |\phi(t)|^2 dt$$

determines a norm for ϕ. Define a projection P on $H(\omega)$ by

$$(P\phi)(t) = \phi_1 \exp(i\omega t) + \phi_{-1} \exp(-i\omega t) ,$$

where ϕ_n is the nth Fourier coefficient of ϕ, $n = \pm 1$. Define $\eta = \{n : R \to R : n(-x) = -n(x)$ and there exists α, β such that $\alpha(x_1 - x_2) \le n(x_1) - n(x_2) \le \beta(x_1 - x_2)$ for all $x_1 < x_2\}$.

Let a_j be real numbers, $0 \le j \le m$ with $a_m \ne 0$ and define $P(D) = \sum_{j=0}^{m} a_j D^m$ where $D = d/dt$. For $n \in \eta$ consider the mth order differential equation

$$P(D)x + n(x) = 0 . \tag{1}$$

If $P(i\omega) \ne 0$ for all $\omega \ge 0$, then for any fixed value of ω, (1) is equivalent to an operator equation

$$x + g\, n(x) = 0 \tag{2}$$

on $H(\omega)$. The operator g can be written as a convolution operator on $H(\omega)$,

$$(g\phi)(t) = \int_{-\infty}^{\infty} G_0(t - s)\phi(s)ds \qquad (G_0 \in L^1),$$

or in terms of Fourier series multiplication

$$(g\phi)(t) \sim \sum_{n\ \mathrm{odd}} G(in\omega)\phi_n \exp(in\omega t) , \tag{3}$$

where $G(\sigma + i\tau)$ is the Laplace transform of $G_0(t)$.

The describing function method as applied to (2) may be described as follows. In (2) replace $n(x)$ by $Pn(x)$, where P is the projection defined above. There results the approximation

$$\overline{x} + Pg\, n(\overline{x}) = 0 . \tag{4}$$

This equation is solved using the describing function

$$N(a) = \frac{1}{\pi a} \int_0^{2\pi} e^{-iu} n(a \cos u) du$$

for the nonlinearity $n(x)$. Solving (4) is equivalent to finding $a \geq 0$ and $b \in [0, 2\pi)$ such that

$$[G(i\omega)N(a) + 1](a/2)\exp(ib) = 0 \text{ or if } a > 0 , \tag{5}$$

$$1 + G(i\omega)N(a) = 0 .$$

One usually attempts to solve (5) graphically.

The major theoretical problems are the following. If (4) has a solution $\bar{x} \in H(\omega)$, does (2) have a nearby solution $x \in H(\omega)$? If (4) has no solution in $H(\omega)$ for all $\omega > 0$, then can we be sure that (2) has no such solution? If (4) has no oscillation in $H(\omega)$ for any $\omega > 0$, then what additional conditions are needed to rule out all periodic and almost periodic oscillations (see [3] for related results).

Analysis of Systems. Consider a system of equations of the form

$$x_k + g_k n_k(x_k) = g_k \sum_{j=1}^{\ell} \gamma_{kj} x_j \quad (k = 1, 2, \ldots, \ell \tag{Σ_k}$$

which could be written in matrix-vector form as

$$x + gn(x) = g\gamma x . \tag{Σ}$$

We assume the following.

$n_k \in \eta$ with constants α_k, β_k where $\alpha_k < \beta_k$, (A1)

g_k is a linear map on $H(\omega)$ defined as in (3) by a continuous function $G_k(i\omega)$ for $\omega \in R$ and the γ_{kj} are real constants, $1 \leq j, k \leq \ell$.

$G_k(i\omega) = \overline{G_k(-i\omega)} \in 0$ for $w \geq 0$ and $G_k(i\omega) \to 0$ (A2)
as $\omega \to 0$ for $k = 1, 2, \ldots \ell$.

Think of (Σ) as the interconnection of the ℓ subsystems

$$x_k + g_k n_k(x_k) = 0 \tag{S_k}$$

with interconnections $g_k \gamma_{kj}$ from (S_j) to (S_k). Define

$$P_k(\omega) = \inf_{n \text{ odd}} \left| \frac{1}{G_k(in\omega)} + \frac{\beta_k + \alpha_k}{2} \right| .$$

Since $G_k(i\omega) \to 0$ as $|\omega| \to \infty$ and $G_k(in\omega) \neq 0$ for all n, then $p_k(\omega) \to \infty$ as $\omega \to \infty$. Define a matrix $R(\omega) = \{r_{kj}(\omega)\}$ as follows:

$$r_{kj}(\omega) = \begin{cases} \dfrac{2}{\beta_k - \alpha_k} p_k(\omega) - 1 - \dfrac{2}{\beta - \alpha} |\gamma_{kk}| & \text{if } k = j \\ \\ \dfrac{-2}{\beta_k - \alpha_k} |\gamma_{kj}| & \text{if } k \neq j. \end{cases}$$

Define $\Gamma = \{\omega > 0 : R(\omega)$ is an M-matrix$\}$.

Theorem 1. If $\omega \in \Gamma$, then (Σ) has no nontrivial solution $x \in H(\omega)$.

Proof. Suppose $x = x_1, \ldots, x_\ell)^T \in H(\omega)$ is a nontrivial solution of (Σ) and that $\omega \in \Gamma$. For a given k the operator $I + (\beta_k + \alpha_k) g_k/2$ has a continuous inverse on $H(\omega)$ if and only if

$$\inf_{n \text{ odd}} \left| 1 + \frac{(\beta_k + \alpha_k)}{2} G_k(in\omega) \right| > 0. \tag{6}$$

Since $R(\omega)$ is an M-matrix, its diagonal elements are positive. Thus, $2p_k(\omega)/(\beta_k - \alpha_k) > 1$ or

$$\inf_{n \text{ odd}} \left| \frac{1 + G_k(in\omega)(\alpha_k + \beta_k)/2}{G_k(in\omega)} \right| = p_k(\omega) > \frac{\beta_k - \alpha_k}{2} > 0.$$

Since $G_k(in\omega) \to 0$ as $|n| \to \infty$, then (6) must be true. Thus, (Σ_k) can be written in the equivalent form

$$x_k = (I + g_k(\beta_k + \alpha_k)/2)^{-1} g_k \left[\frac{\beta_k + \alpha_k}{2} x_k - n_k(x_k) + \sum_{j=1}^{\ell} \gamma_{kj} x_j \right] .$$

Thus,

$$\|x_k\| \le \|(I + g_k(\beta_k + \alpha_k)/2)^{-1} g_k\| \left[\frac{\beta_k - \alpha_k}{2} \|x_k\| + \sum_{j=1}^{\ell} |\gamma_{kj}| \, \|x_j\|\right]$$

$$= P_k(\omega)^{-1} \left[\frac{\beta_k - \alpha_k}{2} \|x_k\| + \sum_{j=1}^{\ell} |\gamma_{kj}| \, \|x_j\|\right]$$

and

$$\left(\frac{2p_k(\omega)}{\beta_k - \alpha_k} - 1\right) \|x_k\| - \sum_{j=1}^{\ell} \frac{2|\gamma_{kj}|}{\beta_k - \alpha_k} \|x_j\| \le 0 .$$

This can be written more compactly as $R(\omega)W \le 0$, where $W = (\|x_1\|, \ldots, \|x_\ell\|)^T$. Since $R(\omega)^{-1} \ge 0$, this means that $W \le 0$, i.e., $\|x_k\| = 0$ for all k. Q.E.D.

In particular Theorem 1 implies that the operator equation (Σ) has no π-symmetric periodic solutions when ω is sufficiently large. (Compare Yorke [7] for similar work for autonomous differential equations.) For small values of ω nothing can be said. For intermediate values of ω, the following result may be applicable.

Consider (Σ_k) with (A1) and (A2) true. Let $N_k(a)$ be the describing function for $n_k(x)$. Note that $\alpha_k \le N_k(a) \le \beta_k$ for all $a \ge 0$. Define

$$\rho_k(\omega) = \inf_{\substack{n > 1 \\ n \text{ odd}}} \left| G_k(in\omega)^{-1} + \frac{\alpha_k + \beta_k}{2} \right| ,$$

$$\theta_k(\omega) = \min_{a \ge 0} \left| N_k(a) + G_k(i\omega)^{-1} \right| ,$$

$$\lambda_k(\omega) = \frac{\beta_k - \alpha_k}{2\rho_k(\omega)}, \quad \zeta(\omega) = \max_k \left(\frac{\sum_{j=1}^{\ell} |\gamma_{kj}|}{\rho_k(\omega)} \right)$$

$$\kappa(\omega) = \max_k \sum_{j=1}^{\ell} \frac{\beta_j - \alpha_j}{\beta_k - \alpha_k} \frac{|\gamma_{kj}|}{\theta_j(\omega)} ,$$

$$\mu_1(\omega) = (1 - \zeta(\omega))^{-1}, \quad \mu_2(\omega) = (1 - \kappa(\omega))^{-1}$$

and

$$\sigma_k(\omega) = \frac{\left[\frac{\beta_k - \alpha_k}{2}\,\mu_1(\omega)\right]^2}{\rho_k(\omega) - \mu_1(\omega)\frac{\beta_k - \alpha_k}{2}}.$$

Define

$$\Gamma_1 = \{\omega > 0 : \zeta(\omega) < 1,\ \theta_k(\omega) > 0,\ \kappa(\omega) < 1 \text{ and}$$
$$1 - \zeta(\omega) > \lambda_k(\omega) \text{ for } 1 \le k \le \ell\}.$$

Theorem 2. If $\omega \in \Gamma_1$ and if for all $a \ge 0$ and all $k = 1, 2, \ldots, \ell$

$$|N_k(a) + G_k(i\omega)^{-1}| > \sigma_k(\omega)[\mu_2(\omega)/\mu_1(\omega)],$$

then (Σ) has no solution $x \in H(\omega)$ with $x \not\equiv 0$.

Proof. Define $\alpha = \mathrm{diag}(\alpha_1, \alpha_2, \ldots, \alpha_\ell)$, $\beta = \mathrm{diag}(\beta_1, \ldots, \beta_\ell)$ and let $P^* = I - P$ be the projection complementary to P. Then (Σ) is equivalent to

$$\bar{x} + Pgn(\bar{x} + x^*) = g\gamma\bar{x} \tag{7}$$

and

$$x^* + P^*gn(\bar{x} + x^*) = g\gamma x^*. \tag{8}$$

Suppose that (7)-(8) has a nontrivial solution. Write (8) as

$$\left(I + P^*g\,\frac{\alpha + \beta}{2}\right)\left(I - \left(I + P^*g\,\frac{\alpha + \beta}{2}\right)^{-1} P^*g\gamma\right)x^*$$
$$= -P^*g\left[n(\bar{x} + x^*) - \frac{\alpha + \beta}{2}(\bar{x} + x^*)\right].$$

Since $\rho_k(\omega) > 0$ for all k, then $\inf\{|1 + G_k(in\omega)(\alpha_k + \beta_k)/2| : n > 1,\ n \text{ odd},$ $k = 1, 2, \ldots, \ell\} > 0$, so that $(1 + P^*g(\alpha + \beta)/2)$ has a bounded inverse on $H(\omega)$.

Define $A_{kj} = (I + P^*g_k(\alpha_k + \beta_k)/2)^{-1}p^*g_k\gamma_{kj}$. Since we are using the maximum component norm on vectors, then the matrix operator $A = \{A_{kj}\}$ will have norm

$$\|A\| = \max_k \sum_{j=1}^{\ell} \|A_{kj}\| = \max_k \sum_{j=1}^{\ell} \rho_k(\omega)^{-1} |\gamma_{kj}|$$

$$= \zeta(\omega) < 1.$$

Thus, $I - A$ has an inverse which satisfies $\|(I - A)^{-1}\| \leq (1 - \zeta(\omega))^{-1} = \mu_1(\omega)$.

We have shown that (8) is equivalent to

$$x^* = Fx^*,$$

$$Fx^* \equiv -(I - A)^{-1}(I + P^*g(\alpha + \beta)/2)^{-1}P^*g[n(\overline{x} + x^*) - \frac{\alpha + \beta}{2}(\overline{x} + x^*)].$$

Moreover, F is Lipschitz with Lipschitz constant

$$L(\omega) \equiv \mu_1(\omega) \max_k \|(I + P^*g_k(\alpha_k + \beta_k)/2)^{-1}P^*g_k\| (\beta_k - \alpha_k)/2$$

$$= \mu_1(\omega) \left[\max_k \lambda_k(\omega)\right] < 1, \tag{9}$$

for any $\omega \in \Gamma_1$. This shows that F is a contraction map and that for each fixed $\overline{x} \in PH(\omega)$ there is a unique x^* which solves (8). Moreover,

$$\|x^*\| \leq \mu_1(\omega) \max_k [\|x_k^* + \overline{x}_k\| (\beta_k - \alpha_k)/(2\rho_k(\omega))]$$

$$\leq \mu_1(\omega) \max_k [\lambda_k(\omega)(\|\overline{x}_k\| + \|x_k^*\|)]$$

$$\leq \mu_1(\omega) \max_k \lambda_k(\omega)(\|\overline{x}\| + \|x^*\|)$$

$$= \mu_1(\omega) \lambda_m(\|\overline{x}_m\| + \|x^*\|)$$

for some integer m, $1 \leq m \leq \ell$. Hence, for this fixed m

$$\|x^*\| \leq \frac{\mu_1(\omega)\lambda_m(\omega)}{1 - \mu_1(\omega)\lambda_m(\omega)} \|\overline{x}_m\|. \tag{10}$$

Equation (7) is restricted to the subspace $PH(\omega)$. On this subspace Pg has an inverse. Hence, (7) can be written as

$$(Pg)^{-1}x + Pn(\bar{x}) - P\gamma\bar{x} = Pn(\bar{x}) - Pn(\bar{x} + x^*)$$

or

$$2(\beta - \alpha)^{-1}(Pg)^{-1}\bar{x} + 2(\beta - \alpha)^{-1}Pn(\bar{x}) - 2(\beta - \alpha)^{-1}P\gamma\bar{x}$$
$$= -2(\beta - \alpha)^{-1}P(n(\bar{x} + x^*) - n(\bar{x})). \qquad (11)$$

This equation is solved by finding the ± 1 Fourier coefficients. In terms of $G(i\omega)$ and $N(a)$ with $a = (a_1, \ldots, a_\ell)^T$ and a_k the amplitude of $\bar{x}_k$ (11) can be written

$$2(\beta - \alpha)^{-1}G^{-1}(i\omega)a + 2(\beta - \alpha)^{-1}N(a)a - 2(\beta - \alpha)^{-1}\gamma a = F(\omega,a)$$

where $2F(\omega,a)$ represents the first Fourier coefficient of $2(\beta - \alpha)^{-1}P(n(\bar{x}) - n(\bar{x} + x^*))$. Thus,

$$F(\omega,a) = 2[(\beta - \alpha)^{-1}(G(i\omega)^{-1} + N(a)) - (\beta - \alpha)^{-1}\gamma]a$$
$$= 2(I - B)(\beta - \alpha)^{-1}(G(i\omega)^{-1} + N(a))a$$

where $B = (\beta - \alpha)^{-1}\gamma(G(i\omega)^{-1} + N(a))^{-1}(\beta - \alpha)$. Compute

$$|Ba| = \max_k \left| \sum_{j=1}^{\ell} \frac{\gamma_{kj}}{\beta_k - \alpha_k} \frac{(\beta_j - \alpha_j)a_j}{(G_j(i\omega)^{-1} + N_j(a_j))} \right|$$

$$\leq \max_k \sum_{j=1}^{\ell} \frac{\beta_j - \alpha_j}{\beta_k - \alpha_k} \left| \frac{\gamma_{kj}}{(G_j(i\omega)^{-1} + N_j(a_j)} \right| \max_k |a_k|$$

$$\leq \max_k \sum_{j=1}^{\ell} \frac{\beta_j - \alpha_j}{\beta_k - \alpha_k} \frac{|\gamma_{kj}|}{\theta_j(\omega)} |a| = \kappa(\omega)|a| < |a|.$$

Hence, $I - B$ is invertiable and

$$|(I - B)^{-1}| \leq (1 - \kappa(\omega))^{-1} = \mu_2(\omega).$$

Thus,

$$2(\beta - \alpha)^{-1}(G(i\omega)^{-1} + N(a))a = (I - B)^{-1}F(\omega,a)$$

and

$$\begin{aligned} |2(\beta - \alpha)(G(i\omega)^{-1} + N(a))a| &\le \mu_2(\omega)|F(\omega,a)| \\ &\le \mu_2(\omega) \max_k |F_k(\omega,a)| \\ &= \mu_2(\omega) \max_k \|(2/(\beta_k - \alpha_k))P(n_k(\bar{x}_k) - n_k(\bar{x}_k + x_k^*))\| \\ &\le \mu_2(\omega) \max_k \|x_k^*\| = \mu_2(\omega) \|x^*\| \\ &\le \mu_2(\omega)\mu_1(\omega)\lambda_m(\mu) \|\bar{x}_m\| (1 - \mu_1(\omega)\lambda_m(\omega))^{-1} \end{aligned}$$

by (10). By the definition of the vector norm

$$\begin{aligned} &|2(\beta - \alpha)^{-1}(G(i\omega)^{-1} + N(a))a| \\ &= \max_k |2(\beta_k - \alpha_k)^{-1}(G_k(i\omega)^{-1} + N_k(a_k))a_k| \\ &\ge 2(\beta_m - \alpha_m)^{-1} |G_m(i\omega)^{-1} + N_m(a_m)| |a_m| \end{aligned}$$

and $|a_m| = \|\bar{x}_m\|$. Therefore,

$$2(\beta_m - \alpha_m)^{-1} |G_m(i\omega)^{-1} + N_m(a_m)| \le \mu_1\mu_2\lambda_m(1 - \mu_1\lambda_m)^{-1}$$

or

$$|G_m(i\omega)^{-1} + N_m(a_m)| \le \frac{(\mu_2/\mu_1)[(\beta_m - \alpha_m)\mu_1/2]^2}{\rho_m - (\beta_m - \alpha_m)\mu_1/2} = (\mu_2/\mu_1)\sigma_m.$$

This contradicts the hypotheses of the theorem. Thus, no such periodic solution of (Σ) can exist. Q.E.D.

Several remarks are in order. First, note that in the limit as $|\gamma| \to 0$ we have $\zeta(\omega) \to 0$ uniformly for ω on compact subsets of $(0,\infty)$. In this limiting case Theorem 2 above reduces to cases 2 and 3 of the Theorem in [4].

The graphical analysis in [4] can be extended to Theorem 2 as follows. To apply Theorem 2 it is necessary that $\omega \in \Gamma_1$. Thus, there is a minimum ω which can be considered, i.e., $\Gamma_1 = (\omega_m, \infty)$ and ω_m is this minimum. By Theorem 1 there is a maximum ω_M above which there can be no solution of (Σ) in $H(\omega)$. Hence, we can work on the bounded set $J = (\omega_m, \omega_M)$. The number $\mu_1(\omega)$ and the ratio $\mu_2(\omega)/\mu_1(\omega)$ can be computed at a given point ω (or at a grid of points) in J. At this point the interconnection terms will not be needed further. We can concentrate on the free subsystems (S_k). For convenience of notation we will drop the subscript k.

For a subsystem

$$x + gn(x) = 0 \qquad (S)$$

compute $\sigma(\omega)$ as follows. In the complex plane plot the locus of points $-1/G(i\omega)$ for $0 < \omega < \infty$. Now fix $\omega > 0$ and define $P_n = -1/G(in\omega)$ for $n = 1, 3, 5, \ldots,$ i.e., the P_k are the points on the locus $-1/G$ which measure the response of the linear system to higher harmonics. Now draw the critical circle, that is the circle centered at $C = (\alpha + \beta)/2$ with radius $r = \mu_1(\beta - \alpha)/2$. Define $P^0 = P_n$ for that value of $n (n > 1,\ n$ odd) which is closest to (but outside of) the critical circle. Then $\sigma(\omega)$ is the length computed as follows using Figure 1. Draw the line segment from P^0 to C and then erect a perpendicular at C. Draw the square with sides of length r which determine the point D. The point E and the length $\sigma(\omega)$ are defined using similar triangles as in Figure 1.

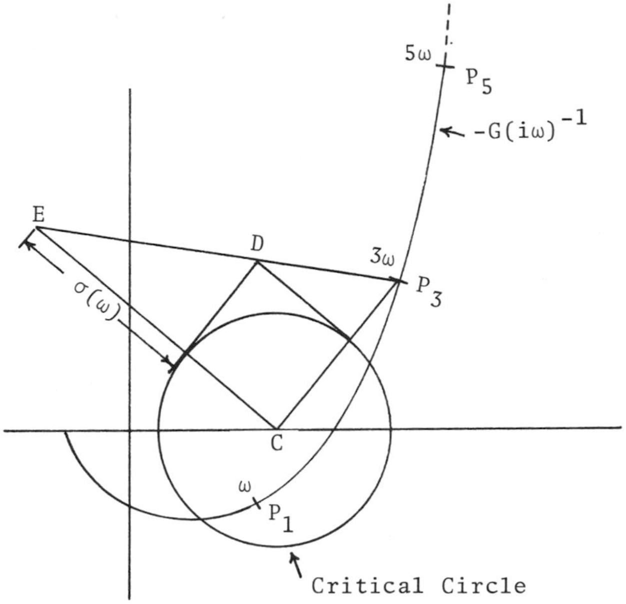

Fig. 1. Finding $\sigma(\omega)$.

We can now specify an uncertainty band, see Figure 2. In the complex plane plot the locus $-1/G(i\omega)$ for $\omega \in J$. For a fixed $\omega \in J$ draw a circle with center at $-1/G(i\omega)$ and radius $\overline{\sigma}(\omega) = \sigma\mu_2/\mu$. The envelope of all such circles for all $\omega \in J$ defines the uncertainty band for the system. The hypotheses of Theorem 2 will be satisfied for all $\omega \in J$ and all $a > 0$ when no part of the uncertainty band intersects the locus of points $N(a)$, $a > 0$. The nonintersection insures that (Σ) has no π-symmetric, nontrivial, periodic solutions. (This graphical analysis will reduce to that [4] in the limit as $|\gamma| \to 0$.)

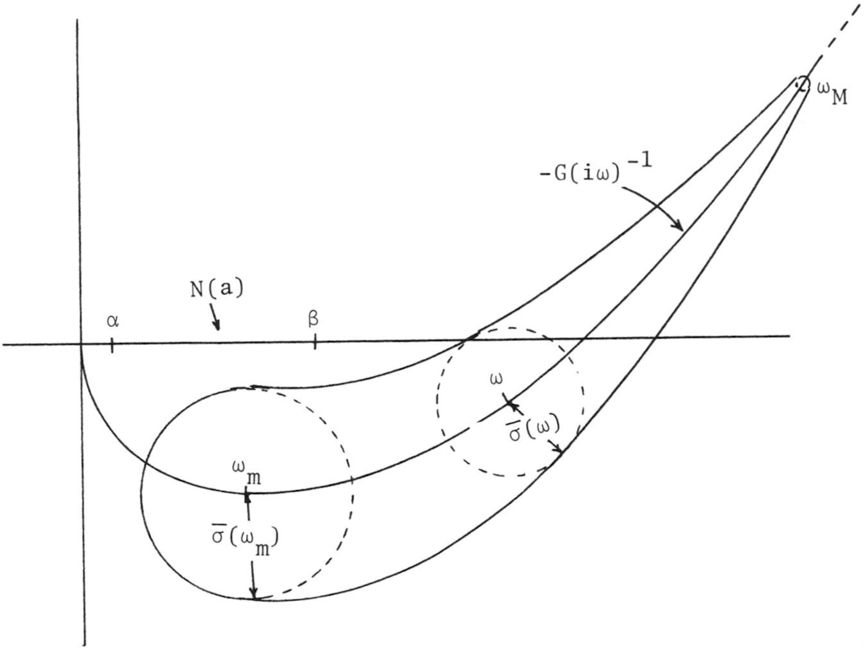

Fig. 2. Uncertainty band.

REFERENCES

[1] Fiedler, M., and V. Ptak, *On matrices with non-positive off-diagonal elements and positive principle minors,* Czechoslovak Math. J., 12(1962), pp. 382-400.

[2] Gelb, A., and W. E. Vonder Velde, "Multiple-Input Describing Functions and Nonlinear System Design," McGraw-Hill, N. Y., 1968.

[3] Loud, W. S., *Nonsymmetric periodic solutions of certain second order nonlinear differential equations,* J. Diff. Eqns., 5(1969), pp. 352-368.

[4] Mees, A. I., and A. R. Berger, *Describing functions revisited,* IEEE Transactions on Automatic Control, AC-20(1975), pp. 473-478.

[5] Michel, A. N., and R. K. Miller, "Qualitative Analysis of Large Scale Dynamical Systems," Academic Press, N.Y., 1977.

[6] Miller, R. K., and A. N. Michel, *On existence of periodic motions in nonlinear control systems with periodic inputs,* "Proceedings of the Functional Differential and Integral Equations Conference," Morgentown, W. Va., June 1979, to appear.

[7] Yorke, J. A., *Periods of periodic solutions and the Lipschitz constant,* Proc. Amer. Math. Soc., 22(1969), pp. 509-512.

BIFURCATION RESULTS FOR EQUATIONS WITH NONDIFFERENTIABLE NONLINEARITIES

Klaus Schmitt

University of Utah

INTRODUCTION

Many physical problems lead to boundary value problems for nonlinear differential equations which depend upon a parameter and which for certain values of the parameter admit multiple solutions. For example, the problem may admit a trivial state as a solution and at certain values of the parameter nontrivial solution branches will bifurcate from the trivial state. To illustrate, let us consider the nonlinear Sturm-Liouville problem

$$Lu = \lambda a(\cdot)u + f(\cdot,u,u',\lambda) \tag{0.1}$$

$$u(0) = 0 = u(1), \tag{0.2}$$

where L is the second order differential operator $(Lu)(x) = -(p(x)u')' + q(x)u$, where $p(\cdot) > 0$, $q(\cdot)$, and $a(\cdot) > 0$ are continuous on $[0,1]$ and $f : [0,1] \times \mathbb{R} \times \mathbb{R} \times \mathbb{R} \to \mathbb{R}$ is continuous and satisfies

$$|f(x,u,v,\lambda)| = o(|u| + |v|) \tag{0.3}$$

as $|u| + |v| \to 0$, uniformly with respect to (x,λ) on compact sets. It is then well known (an application of the Krasnosel'skii-Rabinowitz [11] bifurcation theorem) that each eigenvalue of the linear problem

ISBN 0-12-045550-1

$$Lu = \lambda a(\cdot)\, u \tag{0.4}$$

$$u(0) = 0 = u(1) \tag{0.5}$$

is a bifurcation point for the nonlinear problem (i.e., if $\lambda = \lambda_i$ is an element of $\{\lambda_0, \lambda_1, \ldots\}$ the set of eigenvalues of (0.4), (0.5)), then, the point $(\lambda_i, 0)$ is an accumulation point of solutions (λ, u), $u \neq 0$ (in $\mathbb{R} \times C^1[0,1]$) of (0.1), (0.2)), furthermore, only those points are bifurcation points and from each of these points an unbounded continuum (in $\mathbb{R} \times C^1[0,1]$) of solutions of (0.4), (0.5) branches off. It is also characteristic of Sturm-Liouville problems for o.d.e.'s that such continua cannot bifurcate from other bifurcation points, as follows from elementary nodal properties of solutions of o.d.e.'s and an application of one of the alternatives of Rabinowitz' [11] bifurcation theorem. In this whole development the o-condition (0.3) is a very important one since it implies the Frechet differentiability of the nonlinear part of the abstract operator equation equivalent to (0.1), (0.2). There are, on the other hand, very simple physical situations which lead to nonlinear differential equations problems whose nonlinearities do not satisfy the differentiability condition (0.3). For example, consider the situation of a thin beam which is supported at one end, acted upon by a variable load at the other end, a restoring force on one side and none on the other.

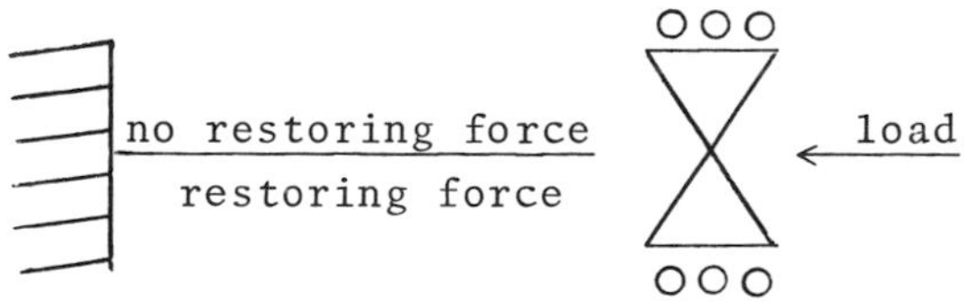

A nonlinear differential equation which describes this situation is

$$\begin{cases} u'' + \lambda(u + g(u,u')) + h(u) = 0 \\ u(0) = 0 = u(1) \end{cases} \tag{0.6}$$

where $g(u,v) = o(|u| + |v|)$ as $|u| + |v| \to 0$ but h is piecewise linear

$$h(u) = \begin{cases} 0 & \text{if} \quad u \le 0 \\ ku & \text{if} \quad u > 0 \end{cases}$$

(if a linear restoring force is acting).

As is to be expected, the bifurcation picture for such a problem, will be quite different from that of a problem like (0.1), (0.2). To illustrate what may happen, we consider two elementary examples.

<u>Example 1</u>. Consider the boundary value problem

$$\begin{cases} u'' + \lambda u + \alpha u^+ + \beta u^- = 0 \\ u(0) = 0 = u(1), \end{cases} \tag{0.7}$$

where $u^+ = \max\{u,0\}$ and $u^- = \max\{-u,0\}$. It is easy to see that the problem will have a solution u with $u(t) > 0$, $t \in (0,1)$ if and only if $\lambda + \alpha = \pi^2$ and a negative solution if and only if $\lambda - \beta = \pi^2$, in each of these cases the problem will have a half ray of solutions emanating from the trivial one, i.e. $\pi^2 - \alpha$ and $\pi^2 + \beta$ are bifurcation points. All other bifurcation points may be computed (see e.g. Fučik [3], where this has been done).

<u>Example 2</u>. Consider the eigenvalue problem

$$\begin{cases} u'' + \lambda g = u \sin \dfrac{1}{\sqrt{\pi^2 u^2 + (u')^2}} \\ u(0) = 0 = u(1). \end{cases} \tag{0.8}$$

We observe that $u(x) = \gamma \sin \pi x$ is a solution if and only if

$$\lambda = \pi^2 + \sin \frac{1}{|\gamma|\pi} ,$$

and thus each $(\lambda,0)$ in $[\pi^2 - 1, \pi^2 + 1] \times C^1[0,1]$ is an accumulation point of nontrivial solutions of (0.8), i.e., bifurcation may take place from a set of points of positive measure.

In the absence of a o-condition on f Bailey [1] already considered the existence theory of nontrivial solutions of (0.1), (0.2) using polar coordinate transformation techniques (Prüfer transformation), this approach has later also been used by Hartman [4] to obtain certain global (with respect to λ) existence results for nontrivial solutions (also for more general and periodic boundary conditions).

In this paper we shall review some general appraoches to the bifurcation problem for operator equations whose nonlinearities are not necessarily differentiable, in this we shall mainly rely on the work of Berestycki [2], MacBain [8], McLeod and Turner [9], Schmitt and Smith [12] and Turner [14]. There is also related work of Keady and Norbury [6] [7] on semilinear Dirichlet problems of the form

$$\begin{cases} \Delta u + \lambda [u - q]_+ = 0 & \text{in } \Omega \subset \mathbb{R}^2 \\ u = 0 & \text{on } \partial\Omega, \end{cases} \tag{0.9}$$

where q is a given positive function.

Work on bifurcation problems whose nonlinearities are not Fréchet differentiable but weakly Fréchet differentiable is due to Wegner [15], who developed an analogue of the Krasnosel'skii-Rabinowitz theorem for such nonlinear equations.

1. GENERAL RESULTS

Consider the operator equation

$$u = F(\lambda,u) \tag{1.1}$$

in a real Banach space E with norm $\|\cdot\|$, where $F : \mathbb{R} \times E \to E$ is completely continuous and satisfies

$$F(\lambda,0) = 0, \quad \lambda \in \mathbb{R}. \tag{1.2}$$

By a solution of (1.1) we mean a pair (λ,u) such that (1.1) holds. A solution is termed nontrivial if $u \neq 0$, and trivial otherwise. Concerning F we shall further assume that for u in a neighborhood of 0

$$F(\lambda,u) \in \lambda Au + H(\lambda,u) + r(\lambda,u), \tag{1.3}$$

where $A : E \to E$ is a compact linear map and $H : \mathbb{R} \times E \to 2^E$ is such that:

$$\text{for each } (\lambda,u), \; 0 \in H(\lambda,u) \tag{1.4}$$

$$\text{for each } (\lambda,u), \; H(\lambda,u) \tag{1.5}$$

is closed, bounded and starlike with respect to $0 \in E$.

$$\text{for each } (\lambda,u), \; H(\lambda,u) \tag{1.6}$$

is positive homogeneous with respect to u, i.e., $\eta H(\lambda,u) \subseteq H(\lambda,\eta u)$, $\eta > 0$.

$$\text{The graph of } H, \; \{((\lambda,u),v) : v \in H(\lambda,u)\} \tag{1.7}$$

is closed in $\mathbb{R} \times E \times E$, and for every bounded set $B \subseteq \mathbb{R} \times E$, $\bigcup_{(\lambda,u) \in B} H(\lambda,u)$ is precompact in E. We shall say that H is completely continuous in this case.

The map $r : \mathbb{R} \times E \to 2^E$ is assumed to satisfy:

$$r(\lambda,u) = o(\|u\|), \quad \|u\| \to 0 \tag{1.8}$$

uniformly on compact λ-sets.

We now associate with (1.1) the following inclusion:

$$u \in \lambda Au + H(\lambda,u) \tag{1.9}$$

and the linear equation

$$u = \lambda A u \tag{1.10}$$

and define

$$\Sigma = \{\lambda \in \mathbb{R} : (1.9) \text{ has a nontrivial solution}\} \tag{1.11}$$

and

$$\Sigma_A = \{\lambda \in \mathbb{R} : (1.10) \text{ has a nontrivial solution}\}. \tag{1.12}$$

Further, we let

$$\mathcal{B} = \{\lambda : (\lambda,0) \ \text{ is a bifurcation point for } (1.1)\}. \tag{1.13}$$

Remark. It will be seen that the positive homogeneity (1.6) in fact only needs to hold for $n \geq 1$.

We use as an example the types of operator equations considered by Turner [14] to illustrate how $\mathcal{H}$ may be constructed.

Assume that $F(\lambda,u)$ is of the form

$$F(\lambda,u) = \lambda Au + G(u)u , \tag{1.14}$$

where for each u, $u \mapsto G(u)$ is a compact linear operator, in this case, we may let

$$\mathcal{H}(u) = \{v : v = sG(tu)u,\ 0 \leq s \leq 1,\ 0 \leq t \leq 1\}. \tag{1.15}$$

We also remark here that Σ may be characterized as follows: (see Szostak [13])

$$\Sigma = \{\lambda \in \mathbb{R} : \inf_{\|y\|=1} \operatorname{dist}(y - \lambda Ay, \mathcal{H}(y)) = 0\}.$$

It is clear that Σ is closed and the following relationship between $\mathcal{B}$ and Σ pertains.

Lemma 1.1. $\mathcal{B} \subseteq \Sigma$. More precisely, if $\lambda \notin \Sigma$, then there exists a neighborhood of $(\lambda,0) \in \mathbb{R} \times E$ containing only the trivial solutions of (1.1) .

PROOF. Suppose $\lambda \in \mathcal{B}$. Then there exists a sequence $\{(\lambda_n,x_n)\}$ such that $(\lambda_n,x_n) \to (\lambda,0)$, $x_n \neq 0$ and $x_n = F(\lambda_n,x_n)$. Hence, there exists $w_n \in \mathcal{H}(\lambda_n,x_n)$,

$z_n \in r(\lambda_n, x_n)$ such that

$$x_n = \lambda_n A x_n + w_n + z_n .$$

Letting $v_n = \frac{x_n}{\|x_n\|}$, we obtain

$$v_n = \lambda_n A v_n + u_n + r_n ,$$

where $r_n \to 0$ and $u_n \in H(\lambda_n, v_n)$. Using the complete continuity of A and H one obtains that $\{u_n\}$, and $\{v_n\}$ have a convergent subsequence say (using a relabelling)

$u_n \to u$, $v_n \to v$, $\|v\| = 1$ and

$$v = A_v + u ,$$

where $u \in H(\lambda, v)$; thus, $\lambda \in \Sigma$.

We now state the analogue of the Krasnosel'skii-Rabinowitz bifurcation theorem for equation (1.1). We use the following notation: S is the closure in $\mathbb{R} \times E$ of $\{(\lambda,u) : u = F(\lambda,u),\ u \neq 0\}$, and, if $\lambda \in \Sigma_A$, then its multiplicity is defined as the dimension of the generalized eigenspace, i.e.,

$$\dim \bigcup_{j=1}^{\infty} \ker(I - \lambda A)^j .$$

Theorem 1.2. Let $a, b \in \mathbb{R} \setminus \Sigma$ and assume that $[a,b]$ contains an odd number of elements of Σ_A , counting multiplicity. Then $\mathcal{B} \cap [a,b] \neq \phi$. Let $\mathcal{C} = \{C : C$ is a subcontinuum of $([a,b] \cap \mathcal{B}) \times \{0\} \cup S$ which meets $([a,b] \cap \mathcal{B}) \times \{0\}$. Then either

(i) $K = \bigcup_{C \in \mathcal{C}} C$ is unbounded.

or

(ii) there exist $C \in \mathcal{C}$ such that C meets $(\mathbb{R} \setminus [a,b]) \times \{0\}$.

The proof of this result is somewhat lenghty, we shall hence not given in its entirety but refer to [12] for the details. One part of the result, however, is easily seen to be a consequence of the homotopy invariance theorem of the Leray-Schauder degree, let us therefore prove here that $\mathcal{B} \cap [a,b] \neq \phi$. To do this, we show that for $\delta > 0$, sufficiently small

$$d_{LS}(I - F(a,\cdot), B_\delta, 0) \neq d_{LS}(I - F(b,\cdot), B_\delta, 0), \tag{1.16}$$

where $B_\delta = \{u \in E : \|u\| < \delta\}$. Let t denote either a or b, we claim that for $\delta > 0$ sufficiently small the LS degrees above are defined and in fact that

$$d_{LS}(I - F(t,\cdot), B_\delta, 0) = d(I - tA, B_\delta, 0), \tag{1.17}$$

the Leray-Schauder formula then tells us that $0 \neq d_{LS}(I - aA, B_\delta, 0) = (-1)^\beta\, d_{LS}(I - bA, B_\delta, 0)$, where β is the sum of the multiplicities of the elements of Σ_A contained in $[a,b]$. To prove (1.17) we consider the homotopy

$$I - \mu F(t,\cdot) - (1 - \mu)tA.$$

That this mapping is zero free on ∂B_δ for $\delta > 0$ small and $0 \leq \mu \leq 1$, follows by an argument similar to that used in the proof of Lemma 2.1. Hence, it follows that for every small δ there exists $\lambda \in (a,b)$ such that (1.1) has a solution u with $\|u\| = \delta$, proving the claim.

We consider some elementary examples from [12] to illustrate the result; more important applications to ordinary and functional differential equations will follow later.

Example. For the scalar equation

$$u = \lambda u + |u| ,$$

we have $\Sigma_A = \{1\}$. $H(\lambda,u)$ may be taken to be $[0,|u|]$, we thus obtain $\Sigma = [0,2]$ and easily compute $\mathcal{B} = \{0,2\}$.

Example. For the scalar equation

$$u = \lambda u + u \sin \frac{1}{u},$$

we may take $H(\lambda,u) = [-|u|, |u|]$, again $\Sigma = [0,2]$ and now $\mathcal{B} = [0,2]$.

Further examples can be constructed to show that $\mathcal{B}$ may in fact be any closed subset of (see again [12]).

As in the differentiable case, both alternatives of Theorem 1.2 may occur. In order to decide which alternative is valid in a given situation the following result is of help, it gives sufficient conditions which present the second alternative from occurring.

Lemma 1.3. Let $\mathcal{O}$ be a nonempty open subset of E such that $t\mathcal{O} \subset \mathcal{O}$, $t > 0$. Assume that (1.1) has no nontrivial solutions (λ,u) with $u \in \partial\mathcal{O}$. Let $a, b \in \mathbb{R}\setminus\Sigma$, $[a,b] \cap \Sigma \neq \phi$, and assume that every nontrivial solution u of (1.9) with $\lambda \in (a,b) \cap \Sigma$ lies in $\mathcal{O}$. Then there exists $\varepsilon_0 > 0$ such that if (1.1) has a solution $(\lambda,u) \in [a,b] \times (B_{\varepsilon_0} \setminus \{0\})$, then $u \in \mathcal{O}$.

The proof of this lemma proceeds by arguments like the ones used in the proof of Lemma 1.1 and Theorem 1.2. For the proof of the next result we again refer to [12].

Theorem 1.4. Let a, b and $\mathcal{O}$ be as in Lemma 1.3, and let there exist another nonempty open set $U \subset E$ such that $tU \subset U$, $t > 0$, $0 \notin U$, $\mathcal{O} \cap U = \phi$, and such that if $u = F(\lambda,u)$, $u \neq 0$, then $u \notin \partial U$, but $u \in \mathcal{O} \cup U$. In addition, assume that if (1.9) has a solution with $u \neq 0$, $\lambda \in (\mathbb{R}\ [a,b]) \cap \Sigma$, then $u \in U$. Then if $\mathcal{B} \cap [a,b]) \neq \phi$

and if C is any subcontinuum of $(\mathcal{B} \cap [a,b]) \times \{0\} \cup S$ meeting $(\mathcal{B} \cap [a,b]) \times \{0\}$, then $C \subset \mathbb{R} \times \mathcal{O} \cup ([a,b] \times \{0\})$.

Remark. It easily follows from Theorems 1.2 and 1.4, that if in fact [a,b] contains only one eigenvalues of A of odd multiplicity, i.e., a single element of Σ_A, then the hypotheses of Theorem 1.4 in fact guarantee that the continuum C must be unbounded.

Let us for the moment assume that each element of Σ_A is of odd multiplicity and assume that for each i there exists $\mathcal{O}_i$, $0 \notin \mathcal{O}_i$, $t\mathcal{O}_i \subset \mathcal{O}_i$, $t > 0$, $\mathcal{O}_i \cap \mathcal{O}_j = \phi$, $i \neq j$, such that the eigensolutions of $u = \lambda_i Au$ belong to $\mathcal{O}_i$. Furthermore, assume that (1.1) has the form (1.14) and that for $0 \leq \varepsilon \leq 1$ nontrivial solutions u of

$$u = \lambda Au + G(u)(|u|^{\varepsilon} u) \tag{1.18}$$

must belong to $\bigcup_{1 \leq i < \infty} \mathcal{O}_i$.

Note if we consider (1.18) for $0 < \varepsilon \leq 1$, then $G(u)(|u|^{\varepsilon}u) = o(\|u\|)$ as $\|u\| \to 0$. Hence, we may apply the Krasnosel'skii-Rabinowitz theorem and conclude that each element of Σ_A is a bifurcation point of (1.18) and in fact that a continuum will bifurcate from λ_i which resides in $\mathcal{O}_i$ and (since $\mathcal{O}_i \cap \mathcal{O}_j = \phi$, $i \neq j$) thus must be unbounded in $\mathbb{R} \times E$. Let us again assume that there exist a, $b \notin \Sigma$ and let $\lambda_i \in (a,b)$. Let $\delta > 0$ be given and let $N_\delta = (a,b) \times B_\delta$. Let $\{\varepsilon_\eta\}$ be a sequence of positive real numbers decreasing to 0 and consider the sequence of problems

$$u = \lambda Au + G(u)|u|^{\varepsilon_n} u.$$

By the arguments above each of these problems will have a solution (λ_n, u_n) with $u_n \in \mathcal{O}_i$ and $u_n \in \partial N_\delta$. Letting

$v_n = \frac{u_n}{|u_n|}$, we see, using by now standard arguments that $\{v_n\}$ has a subsequence which converges to, say, v and v must satisfy (if $u_n \to 0$)

$$v \in \lambda Av + H(v),$$

where λ is the limit of an appropriate subsequence of $\{\lambda_n\}$. Hence, since $v \neq 0$, λ cannot equal a or b. Thus, an appropriate subsequence of $\{u_n\}$ will converge to a nonzero u and u must satisfy

$$u = \lambda Au + G(u)u,$$

furthermore, $u \in \mathcal{O}_i$. Using a somewhat more refined argument one in fact may show that an unbounded continuum C of solutions of (1.14) must bifurcate from $[a,b]$ and $C \subset \mathcal{O}_i$. These types of arguments have been used by Berestycki [2] to treat nonlinear Sturm-Liouville boundary value problems.

Let us assume that $F(\lambda,u)$ satisfies the condition

$$F(\lambda,u) \in \lambda Au + H(\lambda,u) \tag{1.19}$$

for all (λ,u) (we remark here once again that for the proofs of the above results (1.3) was only required near $u = 0$), furthermore, for each λ let there exist a constant $M(\lambda)$ such that

$$v \in H(\lambda,u) \Rightarrow \|v\| \leq M(\lambda)\|u\|. \tag{1.20}$$

Let $\lambda \in \Sigma \setminus \Sigma_A$ and let $u \neq 0$ be such that

$$u \in \lambda Au + H(\lambda,u),$$

then there exists $v \in H(\lambda,u)$ such that

$$u = \lambda Au + v,$$

but since $\lambda \notin \Sigma_A$, we get

$$u = (I - \lambda A)^{-1}v$$

or

$$\|u\| \leq \|(I - \lambda A)^{-1}\| M(\lambda) \|u\| .$$

Let $N(\lambda) = \|(I - \lambda A)^{-1}\|$, we hence conclude that in this case

$$N(\lambda)M(\lambda) \geq 1. \tag{1.21}$$

Using these ideas we obtain the following surjectivity result.

Theorem 1.5. Let (1.19) hold and assume there exists λ such that

$$N(\lambda)M(\lambda) < 1, \tag{1.22}$$

where $N(\lambda)$ and $M(\lambda)$ are given as above. Then for every $v \in E$ there exists a solution u of

$$u = F(\lambda,u) + v. \tag{1.23}$$

PROOF. The result follows from the Leray-Schauder continuation theorem. We consider the family of problems

$$u = \lambda Au + \mu(F(\lambda,u) - \lambda Au) + \mu v, \ 0 \leq \mu \leq 1. \tag{1.24}$$

We note that because of (1.22) we have for $\mu = 0$ the unique solution $u = 0$. It suffices hence to show that solutions of (1.24) are a priori bounded for $\mu \in [0,1]$. If this were not the case, there would exist a sequence $\{u_n\}$, $\|u_n\| \to \infty$ and μ_n such that

$$u_n = \lambda Au_n + \mu_n(F(\lambda,u_n) - \lambda Au_n) + \mu_n v.$$

We let $w_n = \frac{u_n}{\|u_n\|}$ and observe that

$$w_n \in \lambda Aw_n + \mu_n H(\lambda,w_n) + \mu_n \frac{v}{\|u_n\|} .$$

Using the hypotheses on H we get a subsequence of $\{w_n\}$ which converges to, say, w and w must satisfy

$$w \in \lambda Aw + H(\lambda,w)$$

contradicting (1.22).

Remark. The term v in (1.23) may be replaced by $G(u)$ where G is quasi bounded, with quasibound 0, i.e.,

$$\lim_{\|u\| \to \infty} \frac{G(u)}{\|u\|} = 0,$$

the proof is similar in this case.

2. NONLINEAR STURM-LIOUVILLE PROBLEMS

Let L denote the differential operator

$$Lu = -(pu')' + qu \ , \tag{2.1}$$

where p is positive continuous and q is continuous on $[0,\pi]$. The boundary conditions imposed are

$$\begin{cases} b_0 u(0) + c_0 u'(0) = 0 \\ b_1 u(\pi) + c_1 u'(\pi) = 0 \ , \end{cases} \tag{2.2}$$

where $(b_0^2 + c_0^2)(b_1^2 + c_1^2) > 0$.

Let a be a positive continuous function on $[0,\pi]$ and consider the equation

$$Lu = \lambda au + h(t,u,u',\lambda) \tag{2.3}$$

subject to the boundary conditions (2.2). We assume that $h = f + g$ satisfies

$$|f(t,u,v,\lambda)| \le M(\lambda)|u| \tag{2.4}$$

$$|g(t,u,v,\lambda)| = o(|u| + |v|) \tag{2.5}$$

both near $u = v = 0$.

Remark. In [12] the more general Lipschitz condition

$$|f(t,u,v,\lambda)| \le M(\lambda)|u| + K(\lambda)|v|$$

is assumed, since we here wish to avoid too many technical details we restrict ourselves to requirement (2.4).

Assuming that $\lambda = 0$ is not an eigenvalue of (2.3), (2.2) (when $h = 0$), we may rewrite (2.3), (2.2) as the equivalent problem

$$u = \lambda Au + F(\lambda,u) + G(\lambda,u), \tag{2.6}$$

where

$$(Au)(t) = \int_0^\pi G(t,s)a(s)u(s)ds \tag{2.7}$$

$$F(\lambda,u)(t) = \int_0^\pi G(t,s)f(s,u(s),u'(s),\lambda)ds \tag{2.8}$$

and

$$G(\lambda,u)(t) = \int_0^\pi G(t,s)g(s,u(s),u'(s),\lambda)ds , \tag{2.9}$$

where $G(t,s)$ is the Green's function associated with the linear problem, and (2.6) is an equation in $E = C^1[0,\pi]$. We let $r(\lambda,u) = G(\lambda,u)$ and define

$H(\lambda,u) = \{v : v(t) = \int_0^\pi G(t,s)h(s)ds$ where h is measurable on $[0,\pi]$ and $|h(s)| \le M(\lambda)|u(s)|$, a.e., on $[0, \pi]\}$.

It is now not difficult to see that with these definitions the requirements imposed earlier are satisfied.

In this case $\Sigma_A = \{\lambda_0, \lambda_1, \ldots\}$ where $\lambda_0 < \lambda_1 < \ldots < \lambda_n < \ldots$ are the eigenvalues of the linear Sturm-Liouville problem, each of which is of odd multiplicity.

Let us assume that $\lambda \in \Sigma$, then there exists a nonzero u such that

$$u \in \lambda Au + H(\lambda,u),$$

i.e., there exists $v \in H(\lambda,u)$ such that

$$u = \lambda Au + v$$

or there exists a measurable h such that

$$|h(s)| \le M(\lambda)\ |u(s)|$$

$$u(t) = \lambda \int_0^\pi g(t,s)a(s)u(s)ds + \int_0^\pi G(t,s)h(s)ds,$$

or

$$\sqrt{a(t)}\, u(t) = \lambda \int_0^\pi \sqrt{a(t)}\, G(t,s) \sqrt{a(s)} \sqrt{a(s)}\, u(s)ds$$

$$+ \int_0^\pi \sqrt{a(t)}\, G(t,s) \sqrt{a(s)} \sqrt{a(s)}\, \frac{h(s)}{a(s)}\, ds$$

i.e. (letting $\tilde{u} = \sqrt{a}\, u$)

$$\tilde{u} = \lambda H\tilde{u} + H((\frac{\tilde{h}}{a})),$$

where H is the symmetric operator

$$H\tilde{u}(t) = \int_0^\pi \sqrt{a(t)}\; G(t,s)\; \sqrt{a(s)}\; \tilde{u}(s)ds.$$

Hence

$$\tilde{u} = (I - \lambda H)^{-1}H((\frac{\tilde{h}}{a})),$$

or taking L^2 norms we get

$$\|\tilde{u}\|_{L^2} \le \|(I - \lambda H)^{-1}H\|_{L^2}\; \|(\frac{\tilde{h}}{a})\|_{L^2},$$

since H is symmetric

$$\|(I - \lambda H)^{-1}H\|_{L^2} = \frac{1}{\mathrm{dist}(\lambda,\Sigma_H)},$$

thus,

$$\|\tilde{u}\|_{L^2} \le \frac{1}{\mathrm{dist}(\lambda,\Sigma_H)} \left\| \frac{h}{\sqrt{a}} \right\|_{L^2} \le \frac{\frac{M(\lambda)}{a_0}\, \|\tilde{u}\|_{L_2}}{\mathrm{dist}(\lambda,\Sigma_H)},$$

where $a_0 = \min_{[0,\pi]} a(s)$. Thus, it must be the case that

$$\mathrm{dist}(\lambda,\Sigma_H) \le \frac{M(\lambda)}{a_0}. \quad \text{(Note } \Sigma_H = \Sigma_A.\text{)}$$

If, in fact, $M(\lambda) \equiv M$, then one can show, see [12], that

$$\Sigma = \bigcup_{k=0}^{\infty} [\lambda_k - \frac{M}{a_0}, \lambda_k + \frac{M}{a_0}].$$

We summarize these considerations in

Lemma 2.1. Let $\lambda \in \Sigma \setminus \Sigma_A$, then

$$\operatorname{dist}(\lambda, \Sigma_A) \le \frac{M(\lambda)}{a_0}, \tag{2.10}$$

where $a_0 = \min_{[0,\pi]} a(s)$.

If now $M(\lambda) = M$ is independent of λ, then there exists k_0 such that for ℓ, $k \ge k_0$

$$[\lambda_k - \frac{M}{a_0}, \lambda_k + \frac{M}{a_0}] \cap [\lambda_\ell - \frac{M}{a_0}, \lambda_\ell + \frac{M}{a_0}] = \phi$$

and hence for each $k \ge k_0$ $[a,b]$ may be found such that

$$a, b \notin \Sigma, \quad \text{and} \quad [a,b] \cap \Sigma_A = \{\lambda_k\}.$$

Because of the nodal properties of solutions of such a nonlinear Sturm-Liouville problems (solutions may not have multiple zeros) we may choose $\mathcal{O} = S_k = \{u \in C^1 [0,\pi]$: u has only simple zeros and k such in $(0,\pi)\}$ and $U = \bigcup_{\ell \ne k} S_\ell$ and by an easy calculation show that the continua bifurcating from $[a,b]$ must in fact lie in S_k.

If it is the case that $h = f$ and (2.4) holds everywhere with $M = M(\lambda)$, it then follows from arguments like those used in Theorem 1.5 that for k large and λ such that

$$\lambda_k + M < \lambda < \lambda_{k+1} - M$$

and every $v \in L^2(0,\pi)$ there exists a solution of

$$Lu = \lambda a u + f(t,u,u',\lambda) + v$$

which satisfies the boundary conditions (2.2).

As an application of this latter type of alternative let us, by means of a simple example, show how some results of Kanan and Locker may be derived by this alternative.

Let us consider the differential equation

$$Lu = f(t,u,u')u + v \tag{2.11}$$

subject to the boundary conditions (2.2), where

$$\lambda_k < q \le f(t,u,u') \le p < \lambda_{k+1}, \tag{2.12}$$

where λ_k and λ_{k+1} are two consecutive eigenvalues of L. One has the following theorem.

Theorem 2.2. Let (2.12) hold. Then for any $v \in L^2(0,\pi)$, the boundary value problem (2.11), (2.2) has a solution.

PROOF. Let $\tilde{\lambda} = \frac{p+q}{2}$ and consider the family of problems

$$Lu = \lambda u + (f(t,u,u')u - \tilde{\lambda}u) + v \tag{2.13}$$

subject to (2.2). If suffices to show that $\lambda = \tilde{\lambda}$ does not belong to the set Σ of equation (1.9) associated with $Lu = \lambda u + (f(t,u,u')u - \tilde{\lambda}u)$. Since $|f(t,u,u')u - \tilde{\lambda}u| = |f(t,u,u') - \lambda||u|$, the constant M of this problem is given by

$$M = \sup|f(t,u,u') - \tilde{\lambda}|,$$

on the other hand since (2.12) holds, $M \le \frac{p-q}{2}$. Hence $\lambda_k + M < q + \frac{p-q}{2} = \tilde{\lambda} = p - \frac{p-q}{2} < \lambda_{k+1} - M$. Thus, (2.13), (2.2) has a solution for $\lambda_k + M < \lambda < \lambda_{k+1} - M$, and hence in particular for $\lambda = \tilde{\lambda}$.

3. EIGENVALUE PROBLEMS FOR DELAY EQUATIONS

In this section we shall consider Sturm-Louiville problems for linear delay-differential equations and show how the existence and location of eigenvalues and eigenfunctions may be deduced from the results and considerations of the previous sections. Such problems have been studied in great detail by Norkin [10, Chapter III]. To allow immediate comparison we shall consider problems as formulated there but also observe that more general equations (even nonlinear ones could be considered).

Thus, let $\Delta(t)$, $0 \le t \le \pi$, be a nonnegative continuous function and let $M(t)$ be continuous.

We consider the Sturm-Liouville problem

$$u''(t) + \lambda u(t) + M(t)u(t - \Delta(t)) = 0 \tag{3.1}$$

with the boundary conditions

$$u(\pi) = 0,\ u(t) \equiv 0,\ t \le 0, \tag{3.2}$$

we equally well could treat more general conditions

$$\begin{cases} u(0)\cos\alpha + u'(0)\sin\alpha = 0 \\ u(\pi)\cos\beta + u'(\pi)\sin\beta = 0 \\ u(t - \Delta(t)) = u(0)\phi(t - \Delta(t)),\ \text{if}\ \ t - \Delta(t) < 0, \end{cases} \tag{3.3}$$

where $\phi(t)$ is a continuous function defined on the initial set $E_0 = \{s : s = t - \Delta(t),\ 0 \le t \le \pi\}$ with $\phi(0) = 1$, $|\phi(s)| \le 1$, $s \in E_0$.

Assuming, for simplicity, that $\lambda = 0$ is not an eigenvalue of u'' subject to (3.2) we may rewrite (3.1) as an operator equation in $E = C^1[0,\pi]$,

$$u = \lambda Au + F(u),$$

where

$$(Au)(t) = \int_0^\pi G(t,s)u(s)\,ds$$

and

$$(Fu)(t) = \int_0^\pi G(t,s)M(s)u(s - \Delta(s)\,ds,$$

where u satisfies (3.2).

We define

$$H(u) = \{v : v = F(\mu u),\ 0 \le \mu \le 1\}.$$

Let $\lambda \in \Sigma$, then as in the previous section, we compute that

$$\lambda \in \bigcup_{k=1}^{\infty} [k^2 - M_0,\ k^2 + M_0],$$

where $M_0 = \max_{[0,\pi]} |M(t)|$, since now $\Sigma_A = \{1, 4, \ldots\}$.

Thus, one may deduce from the earlier considerations the following result.

Theorem 3.1. There exists k_0 such that if $k \ge k_0$, then the boundary value problem (3.1), (3.2) has an eigenvalue in each of the intervals $[k^2 - M_0, k^2 + M_0]$, where $M_0 = \max_{t \in [0,\pi]} |M(t)|$. In particular, k_0 is the smallest integer such that $M_0 \le k_0 + \frac{1}{2}$. Furthermore, since $F(u) \in H(u)$ for all u, it follows that the continua bifurcating from an interval $[k^2 - M_0, k^2 + M_0]$ cannot connect up to another such interval, as long as $k > k_0$.

Remark. The estimates used above are easily seen to improve some of the results of Norkin, see in particular the results of chapter 3, sec. 4 of [10].

Since, in general, solutions of second order delay differential equations may have multiple zeros one may not deduce results concerning the oscillatory behavior of eigenfunctions, like the ones in the previous chapter. In particular instances, however, one can. There are simple conditions, e.g., requirements which only depend on the size of M_0 which imply that nontrivial solutions $x(t)$ which satisfy $x(t) \equiv 0$, $t \le 0$, may only have simple zeros (see e.g. Norkin [10, chapter III]). These conditions then imply the same for solutions of $u \in \lambda Au + H(u)$. Thus, in this case one may employ reasoning similar to the one used in the previous case to deduce the existence of an infinite sequence of eigenvalues $\lambda_0 < \lambda_1 < \ldots$ each of which is simple, having associated eigenfunctions which have precisely i nodes interior to $[0,\pi]$.

Remark. To establish the existence of such eigenvalues and eigenfunctions one could of course also employ perturbation theory as developed in Kato [5], however, the results developed here, are equally valid in case $M(t)u(t - \Delta(t))$ is replaced by a nonlinear term $H(t,u(t - \Delta(t))$ which is such that $h(t,0) \equiv 0$ and satisfies a Lipshitz condition $|h(t,v)| \leq M(t)|v|$.

ACKNOWLEDGEMENTS

Some of the work in this paper was done while the author was a U.S. Senior Scientist supported by the Alexander von Humboldt foundation. Their support and the hospitality of the University of Würzburg are gratefully acknowledged.

REFERENCES

[1] Bailey, P., *An eigenvalue theorem for nonlinear second order differential equations*, J. Math. Anal. Appl. 20(1967), pp. 94-102.

[2] Berestycki, H., *On some nonlinear Sturm-Liouville problems*, J. Diff. Equations, 26(1977), pp. 375-390.

[3] Fucik, S., *Boundary value problems with jumping non-linearities*, Cas. Pest. Mat. 101(1976), pp. 69-87.

[4] Hartman, P., *On boundary value problems for suplinear second order differential equations*, J. Diff. Equations, 26(1977), pp. 37-53.

[5] Kato, T.,"Perturbation Theory for Linear Operators," 2nd ed., Springer Verlag, New York, 1976.

[6] Keady, G. and J. Norbury, *A semilinear elliptic eigenvalue problem I,* to appear.

[7] Keady G. and J. Norbury, *A semilinear elliptic eigenvalue problem, II, The plasma problem,* to appear.

[8] MacBain, J. A., *Local and global bifurcation from normal eigenvalues,* Pac. J. Math., 63(1976), pp. 445-466.

[9] McLeod, J. B. and R.E.L. Turner, *Bifurcation for nondifferentiable operators with an application to elasticity,* Arch. Rat. Mech. Anal. 63(1976), pp. 1-45.

[10] Norkin, S. B.,"Differential Equation of the Second Order with Retarded Argument," Transl. Math. Monogr., Vol. 31, Amer. Math. Soc., Providence, 1972.

[11] Rabinowitz, P. H., *Some global results for nonlinear eigenvalue problems,* J. Func. Anal. 7(1971), pp. 487-513.

[12] Schmitt, K. and H. Smith, *On eigenvalue problems for nondifferentiable mappings*. J. Diff. Equations, to appear.

[13] Szostak, E., *Note on an implicit function theorem in a nondifferentiable case,* Zesz. Nank, Univ. Jagiell. Prace Mat. Zesz. 18(1977), pp. 159-162.

[14] Turner, R. E. L., *Nonlinear eigenvalue problems and applications to elliptic equations,* Arch. Rat. Mech. Anal. 42(1971), pp. 184-193.

[15] Wugner, B., *Ein globales Ergebnis für Bifurkationsaufgaben mit schwach differenzierbaren Operatoren*, Math. Machr. 87(1979), pp. 7-14.

THE STRUCTURE OF LIMIT SETS: A SURVEY[1]

George R. Sell
University of Minnesota

INTRODUCTION

Let us consider ordinary differential equations of the form

$$x' = f(x,t) \tag{1}$$

where $x \in R^n$, $t \in R$ and f is continuous and quasi-periodic in t. The quasi-periodicity of f is equivalent to the following: There is an integer $k \geq 0$, a continuous function $F : R^n \times T^k \to R^n$ (where T^k is the k-dimensional torus) a constant vector $\alpha = (\alpha_1, \ldots, \alpha_k) \in R^k$ and a point $\omega \in T^k$ such that $f(x,t) = F(x, \omega \cdot t)$ for all $(x,t) \in R^n \times R^1$, where $\omega \cdot t = \omega + \alpha t \pmod 1$. There is no loss in generality in assuming that the entries in α are linearly independent over the integers, i.e. the only solution of

$$n \cdot \alpha = n_1\alpha_1 + \ldots + n_k\alpha_k = 0 ,$$

where $n_1,\ldots,n_k$ are integers, is $n_1 = \ldots = n_k = 0$. Indeed, if this were not the case, then one can represent f in the form $f(x,t) = \hat{F}(x,\hat{\omega} \cdot t)$ where $\hat{F} : R^n \times T^\ell \to R^n$ (as above) and $\ell < k$.

[1]This research was supported in part by NSF Grant MCS 79-01998

ISBN 0-12-045550-1

The mapping $(\omega,t) \to \omega \cdot t$ describes a flow on T^k. This is sometimes referred to as an "irrational twist flow".

The problem of studying solutions of (1) is then equivalent to studying those of

$$x' = F(x,\omega \cdot t).$$

In the latter case, we will now treat ω as a parameter. We will assume further that F is a C^1-function in x. Consequently, the initial value problem

$$x' = F(x, \omega \cdot t) \ , \ x(0) = x \tag{2}$$

admits a unique solution for every $x \in R^n$ and $\omega \in T^k$. We shall let $\varphi(x,\omega,t)$ denote the noncontinuable solution of (2). Thus, one has $\varphi(x,\omega,0) = x$. Furthermore, the mapping

$$\pi(x,\omega,t) = (\varphi(x,\omega,t) \ , \ \omega \cdot t) \tag{3}$$

is a skew-product flow on $R^n \times T^k$, cf. [27]. We should emphasize that the flow given by (3) may only be "local" flow, since the solutions $\varphi(x,\omega,t)$ may fail to be defined for all $t \in R$.

The study of periodic ordinary differential equations is included in the above context. In this case one has $k = 1$ and $T^k = S^1$. Autonomous equations are also included above with $k = 0$ and where T^o is the set consisting of a single point p with $p \cdot t = p$ for all $t \in R$.

The central problem we wish to study in this paper is the following: Let $\varphi(x_o,\omega_o,t)$ be a solution of (2) that is bounded for $t \geq 0$. Next let

$$\Omega = \Omega(x_o,\omega_o) = \bigcap_{\tau \in R} C\ell\{\pi(x_o,\omega_o,\tau + t) : t \geq 0\}$$

be the ω-limit set of the corresponding motion $\pi(x_o,\omega_o,t)$

in $R^n \times T^k$. The problem is then to classify or describe the various sets Ω that can arise as ω-limit sets.

In this very general setting one can say that Ω is nonempty, compact and connected [4,21] and chain recurrent [10]. In addition if $p : R^n \times T^k \to T^k$ is the natural projection, then one can assert that $p^{-1}(\omega) \cap \Omega$ is a nonempty compact set for every $\omega \in T^k$. In order to get further information about the structure of Ω, one must impose further conditions on either the ambient space R^n or the given solution $\varphi(x_0,\omega_0,t)$. Let us first review some contributions to this problem in the case of autonomous equations.

II. AUTONOMOUS EQUATIONS

For an autonomous differential equation one has $k = 0$ and the functions f and F depend only on $x \in R^n$. In this setting a description of the structure of Ω depends heavily on the dimension n.

If $n = 1$, then $\Omega = \{x_p\}$ contains exactly one point and this point is a fixed point for the given differential equation $x' = f(x)$. (Recall that Ω is the ω-limit set of a bounded solution.)

The first nontrivial result occurs for autonomous differential equations in R^2. The pioneering work of Poincare in 1881-1886 and the later work of Bendixson in 1901 represent one of the first major steps in the qualitative theory of differential equations [3,22]. They showed that if Ω does not contain a fixed point for the given differential equation, then Ω consists of one periodic orbit. (Also, see Hartman [17] for additional information and a more detailed historical commentry.)

One can also replace the ambient space, which is a Euclidean space R^n, with a smooth manifold. A study of the qualitative properties of differential equations on compact 2-dimensional manifolds was initiated by Poincaré [22]. One of the early milestones in the study of differential equations on the 2-dimensional torus T^2 is the paper of Denjoy [11] in 1932. Among other things he showed that there does exist a C^1-vector field on T^2 which has a minimal set Ω that is neither a fixed point, a periodic orbit and $\Omega \neq T^2$. In 1963 Schwartz [29] showed that if Ω is a minimal set for any C^2-vector field on a compact 2-dimensional manifold M then either Ω is a fixed point or periodic orbit, or $\Omega = M = T^2$.

When one increases the dimension of the ambient space, then the ω-limit sets can become extremely complicated. For example, by using the construction of Schweitzer [30] of a C^1-vector field on a solid torus without periodic orbits, we are able to construct a C^1-vector field on R^3 with the following properties:

i) All solutions are bounded.

ii) There are no fixed points, no periodic orbits and no almost periodic solutions.

iii) Every bounded set in R^3 meets at most finitely many minimal sets.

Even greater complexity arises when the dimension $n \geq 3$ with the occurance of strange attractors, see [26, 33, 35] for example. We see then that in these higher dimensional spaces, additional hypotheses are needed in order to show that Ω has some simple structure. Along these lines let us recall the theorem of Birkhoff [5] in 1927. He showed that Ω is a compact minimal set if and only if the flow on Ω is recurrent. Then in 1933, Markov [19] showed that recurrance plus Lyapunov stability (or equicontinuity) implies that Ω is an almost periodic minimal set. This theorem of Markov can be viewed as an early ancestor of the theory of quasi-periodic equations, which we describe next.

III. QUASI-PERIODIC EQUATIONS

The study of quasi-periodic functions began with the work of Bohl [7] in 1893. The extension of Bohl's theory to the larger class of almost periodic functions was given by Bohr [8] in 1925. Then in 1933, Favard [14] wrote his treatise on almost periodicity with applications to linear differential equations with almost periodic coefficients. Favard proved the following result: Assume that the linear inhomogeneous equation

$$x' = A(t)x + f(t) \tag{4}$$

has a bounded solution. Assume further that for every $\tilde{A}$ in the hull $H(A)$ the linear homogeneous equation

$$x' = \tilde{A}(t)x , \tag{5}$$

has the Favard property, which says that for every nontrivial bounded solution $\varphi(t)$ of (5) there is an $\alpha > 0$ such that $|\varphi(t)| \geq \alpha$ for all $t \in R$. Then it follows that (4) has an almost periodic solution. This result also applied to equations with quasi-periodic coefficients. We shall have more to say about this below.

These results by Favard for linear ordinary differential equations have been extended to more general equations including partial differential equations [2,6], differential-delay equations [18] and differential equations in Hilbert spaces [37]. However, we want to redirect our attention now to nonlinear equations in finite dimensional spaces.

In 1955 Amerio [1] gave an important extension of Favard's theory to nonlinear differential equations. He introduced a concept of separatedness for solutions of (1) and then showed that if a bounded solution of (1) is separated then it is almost periodic. In 1965 Seifert [31] showed that global asymptotic stability implied separatedness. Therefore, if (1) had a solution that was bounded for $t \geq 0$ and globally asymptotic stable, then (1) has an almost periodic solution.

The Amerio separatedness condition for the linear equation (4) is somewhat stronger than, but closely related to, the Favard property. In 1972 Fink [15] extended Amerio's Theorem by using a somewhat weaker separatedness condition. He also showed a connection between the separatedness condition and uniform stability. We have more to say about his shortly.

By using different methods based on topological dynamics, Miller [20] showed in 1965 that if the given bounded solution of (2) is uniformly asymptotically stable then (2) has an almost periodic solution. In 1969 Yoshizawa [36] extended Miller's result to differential equations lacking uniqueness by using instead the concept of stability under disturbances.

The Miller-Yoshizawa results are similar in spirit to the Markov Theorem cited above. Also these results generalize Seifert's Theorem because they rely on a local stability property rather than global stability.

At this stage one then had two methods for showing the existence of almost periodic solutions. One method based on the Amerio separatedness condition and the other based on various topological dynamical properties on stability. For a time it was felt that these were distinct methods with only minor overlap. However, in 1977 Sacker and Sell [27], by using the notion of distality in flows, showed that the Amerio-Seifert-Fink theory and the Miller-Yoshizawa theory were both consequences of a single dynamical principle. In order to show the connections between these theories we will need the following definition.

Let Ω be a compact invariant set in the skew-product flow (3) on $R^n \times T^k$. We shall say that the flow π is _distal_ on Ω if for any two points (x_1,ω), $(x_2,\omega) \in \Omega$ with $x_1 \neq x_2$, there is an $\alpha > 0$ such that $|\varphi(x_o,\omega,t) - \varphi(x_2,\omega,t)| \geq \alpha$ for all $t \in R$.

Now let Ω be the ω-limit set of a given bounded solution $\varphi(x_o,\omega_o,t)$ of (2). Then the following items were proved in [27]:

1. If $\varphi(x_o,\omega_o,t)$ is uniformly stable, then the flow π is distal on Ω.

2. Ω is an N-fold covering space of T^k if and only if the flow π is distal on Ω and for some $\omega_1 \in \Omega$ the fibre $\Omega(\omega_1) = \{x \in R^n : (x,\omega_1) \in \Omega\}$ contains exactly N points.

3. If Ω is an N-fold covering space of T^k, then Ω is an almost periodic minimal set and for every $(w,\omega) \in \Omega$ the solution $\varphi(x,\omega,t)$ is almost periodic.

In order to apply this to Amerio's theory one observes that the Amerio separatedness condition implies that Ω is a 1-fold covering space of T^k. For Miller's theory one uses item 1 to conclude that the flow π is distal on Ω. Next one uses the uniform asymptotic stability and the distality to show that fibre $\Omega(\omega_1) = \{x \in R^n : (x,\omega_1) \in \Omega\}$ contains a finite number of points, say N. Hence Ω is an N-fold covering space of T^k by item 2. The almost periodicity is both theories now follows from item 3.

The Favard Theorem for linear equations also fits into this context. The Favard property is, in fact, a statement of distality. Favard's argument consists in show that there is a $\beta \geq 0$ and precisely one bounded solution $\tilde{\varphi}(t)$ of (4) with the following three properties:

i) For any bounded solution $\varphi(t)$ of (4) one has $\beta \leq \sup_t |\varphi(t)|$.

ii) $\beta = \sup_t |\tilde{\varphi}(t)|$.

iii) If $\varphi(t)$ is any bounded solution of (4) with $\beta = \sup_t |\varphi(t)|$ then $\varphi(t) = \tilde{\varphi}(t)$.

Now let Ω be the ω-limit set of this distinguished solution of $\tilde{\varphi}(t)$. From the uniqueness statement (iii), together with the distality of the flow, it is easy to see that Ω is a 1-cover of T^k.

IV. DISTAL PROPERTIES

The theory of Amerio-Miller and others shows that if the given solution $\varphi(x_0,\omega_0,t)$ is uniformly asymptotically stable, then Ω is an N-fold covering space of the base T^k and the flow on Ω is almost periodic. The next step in our program of describing Ω is to weaken the stability assumptions on the solution $\varphi(x_0,\omega_0,t)$.

In 1977 Sacker and Sell [27] showed that if $\varphi(x_0,\omega_0,t)$ is bounded for $t \geq 0$ and uniformly stable, then Ω is a distal minimal set.[2] This result then opens the door to the use of the rather well-developed theory of distal flows (see Ellis [12], Furstenberg [16] and Veech [34]) to the problem of describing Ω. Since Ω lies in the finite dimensional space $R^n \times T^k$, the set Ω is finite dimensional.

A theory of the structure of distal minimal sets of finite topological dimension can be found in Ellis [13], Bronstein [9] and Rees [23,24]. A fairly complete classification is possible when Ω is a manifold. The next question that arises is: when is Ω a manifold? It is possible to give a sufficient condition for this question in terms of the spectral theory developed in Sacker and Sell [28].

With Ω , as given above, we fix $(x,\omega) \in \Omega$ and let $\varphi(t) = \varphi(x,\omega,t)$. Next define the $(n \times n)$ matrix valued function

[2]This result extends to differential-delay equations with almost-periodic coefficients. In this case the solutions generate a semi-flow instead of a flow. However, one can prove that the semi-flow on Ω extends to a two-sided flow, cf. [27, pp. 35-37].

$$A(t) = \frac{\partial}{\partial x} F(x,\omega \cdot t)\bigg|_{x = \varphi(t)} .$$

Now consider the shifted linear equation

$$x' = (A(t) - \lambda I)x , \tag{5}$$

where $\lambda \in R$ is a parameter. Recall that the resolvent set of $x' = A(t)x$ is the set

$$\rho = \{\lambda \in R : (5) \text{ has an exponential dichotomy}\}$$

and the spectrum Σ is the complement $\Sigma = R - \rho$. For $\lambda \in \rho$ let N_λ denote the dimension of the stable manifold of the induced exponential dichotomy. In 1978 Sacker and Sell [28] showed that the spectrum is the union of p non-overlapping compact intervals where $1 \le p \le n$.

It follows from Ellis [13] and Rees [23] that the fibres $\Omega(\omega)$ have constant dimension for $\omega \in T^k$. Let $\ell = \dim \Omega(\omega)$. One then has

$$\dim \Omega = k + \ell .$$

The following theorem is proved in Rees and Sell [25]:

Theorem. The following statements are valid:

(A) If $\mu,\lambda \in \rho$ and $\mu < 0 < \lambda$, then $N_\lambda - N_\mu \ge \ell$.

(B) (Hyperbolic case) If $\mu,\lambda \in \rho$ with $\mu < 0 < \lambda$ and $N_\lambda - N_\mu = \ell$, then Ω is a manifold.

The above theorem generalizes a result of Sell [32] who studied the corresponding problem for autonomous differential equations. In this setting the Markov Theorem is applicable and one knows, in addition, that Ω is an almost periodic minimal set. In the hyperbolic case of the above theorem one can conclude that $\Omega = T^\ell$, an ℓ-dimensional torus.

Let us now look at some examples:

Example 1. Let T^k be a k-dimension torus with a given irrational twist flow $(\omega,t) \to \omega \cdot t$. Let $A : T^k \to R$ be a continuous real-valued function. Consider the corresponding flow on $R^2 \times T^k$ generated by

$$z' = i\,A(\omega \cdot t)z$$

where we identify a point in R^2 with the complex number z in the usual way. Then

$$M = T^1 \times T^k = \{(x,\omega) : |z| = 1\}$$

is an invariant set in this flow and the flow is distal on M. By Ellis' Theorem [12] M is the union of minimal sets. Let Ω denote a minimal set in M . There are now three cases we wish to study:

Case 1. $A(\omega) = \beta$ (a constant) where β is independent of the twist flow on T^k . In this case one necessarily has $\Omega = T^{k+1}$ and the flow on Ω is almost periodic.

Case 2. $\int_0^t [A(\omega \cdot s) - \beta]ds$ is bounded in t , where $\beta = M(A)$ is the mean value of A. In this case one has $\Omega = T^{k+1}$ or Ω is homeomorphic to T^k. Moreover, the flow on Ω is almost periodic.

Case 3. $\int_0^t [A(\omega \cdot s) - \beta]ds$ is unbounded in t, where $\beta = M(A)$. In this case one necessarily has $M = \Omega = T^{k+1}$ and (by Bohr's Theorem) the flow on Ω is not almost periodic.

Notice that in these three cases the fibres have the form

$$\Omega(\omega) = S^1 \quad \text{or} \quad \Omega(\omega) = \{\text{finite set}\} .$$

It is easy to modify this example to find illustrations where $\Omega(\omega) = T^{\ell}$, an ℓ-dimensional torus. In the following example of M. Rees, we describe a distal minimal set where the fibre $\Omega(\omega)$ is S^2, the 2-sphere. The details can be found in [25].

Example 2. Let T^2 be the 2-dimensional torus with a twist flow

$$(\omega_1, \omega_2) \cdot t = (\omega_1 + t, \omega_2 + \alpha t)$$

where α is irrational. Now let M^3 denote the collection of real (3×3) matrices. Then for any continuous matrix valued function $A : T^2 \to M^3$ we let π be the flow on $R^3 \times T^2$ induced by

$$x' = A(\omega \cdot t)x \, , \quad x(0) = x$$

where $x \in R^3$. If the matrix $A(\omega)$ is skew-symmetric for every $\omega \in T^2$, then the set

$$M = \{(x,\omega) : |x| = 1\} = S^2 \times T^2$$

is an invariant set for the flow π. One now has the following result:

Theorem. There is an irrational number α and a C^∞-function $A : T^2 \to M^3$ such that $A(\omega)$ is skew-symmetric for every $\omega \in T^2$ and furthermore the set $M = S^2 \times T^2$ is a distal minimal set for the induced flow π on $R^3 \times T^2$.

REFERENCES

[1] Amerio, L., *Soluzioni quasi-periodiche, o limitate, di sistemi differenziale non-lineari quasi-periodiche, o limitati*. Ann. Mat. Pura Appl. (4) 39(1955), pp. 97-119.

[2] Amerio L., and G. Pruose, "Almost Periodic Functions and Functional Equations." Van Nostrand Reinhold, New York, 1971.

[3] Bendixson, I., *Sur les courbes définiés' per des équations différentielles*, Acta Math. 24(1901), pp. 1-88.

[4] Bhatia, N. P. and G. P. Szegö, "Stability Theory of Dynamical Systems," Springer-Verlag, New York, 1970.

[5] Birkhoff, G. D., "Dynamical Systems," Amer. Math. Soc. Colloquium Publ., New York, 1927.

[6] Bochner, S., *A new approach to almost periodicity*, Proc. Nat. Acad. Aci. U.S.A. 48(1962), pp. 2039-2043.

[7] Bohl, P., *Über die Darstellung von Funktionen einer Variabeln durch trigonometrische Reihen mit mehreren einer Variabeln proportionalen Argumenten*, Dissertation, Dorpat, 1893.

[8] Bohr, H., *Zur Tbeorie der fastperiodische Funktionen*, Acta Math. 46(1925).

[9] Bronstein, I. U., *The structure of distal extensions and finite dimensional distal minimal sets*, Nat. Issled 6(1971), pp. 41-62.

[10] Conley, C. C., "Isolated Invariant Sets and the Morse Index." CBMS Regional Conference No. 38, Amer. Math. Soc., Providence, RI, 1978.

[11] Denjoy, A., *Sur les courbes definies par les équations différentielles à la surface du tore*, J. Math Pures Appl. (9) 11(1932), pp. 33-375.

[12] Ellis, R., *Distal transformation groups*, Pacific J. Math. 8(1958), pp. 401-405.

[13] Ellis, R., "Lectures on Topplogical Dynamics." Benjamin, New York, 1969.

[14] Favard, J. "Lecous sur les Fonctions Presque-Periodique." Gauthier-Villars, Paris, 1933.

[15] Fink, A. M., *Semi-separated conditions for almost periodic solutions*, J. Differential Eqs. 11(1972), pp. 245-251.

[16] Furstenberg, H., *The structure of distal flows*, Amer. J. Math. 85(1963), pp. 477-515.

[17] Hartman, P., "Ordinary Differential Equations," Wiley, New York, 1964.

[18] Kato, J., *Favard's separation theorem in functional differential equations*, Funk Ekvac. 18(1975), pp. 85-92.

[19] Markov, A. A., *Stabilität im Liapunoffschen Sinne und Fastperiodizität*, Math. Z. 36(1933), pp. 708-738.

[20] Miller, R. K., *Almost periodic differential equations as dynamical systems with applications to the existence of a.p. solutions*, J. Differential Eqs. 1(1965), pp. 337-345.

[21] Nemytskii, V. V., and V. V. Stepanov, "Qualitative Theory of Differential Equations," Princeton Univ. Press, Princeton, N. J. 1960.

[22] Poincaré, H., *Memoire sur les courbes défines par les équations différentielle, I, II, III, and IV*, J. Math Pures Appl. (3) 7(1881), pp. 375-422; (3) 8(1882), pp. 251-286; (4) 1(1885), pp. 167-244; (4) 2(1886), pp. 151-217.

[23] Rees, M., *On the fibres of a minimal distal extension of a transformation group.* (to appear).

[24] Rees, M., *On the structure of minimal distal transformations groups with topological manifolds as phase spaces.* (to appear).

[25] Rees, M., and G. R. Sell, *On the structure of limit sets of stable solutions.* (to appear).

[26] Ruells, *Dynamical systems with turbulent behavior,* I.H.E.S. Technical Report, October, 1977.

[27] Sacker, R. J., and G. R. Sell, "Lifting Properties in Skew-Product Flows with Applications to Differential Equations." Memoir. Amer. Math. Soc. No. 190, Providence, RI, 1977.

[28] Sacker, R. J., and G. R. Sell, *A spectral theory for linear differential systems,* J. Differential Eqs., 27(1978), pp. 320-358.

[29] Schwartz, A. J., *A generalization of a Poincare-Bendixson theorem to closed two-dimensional manifolds.* Amer. J. Math. 85(1963), pp. 453-458.

[30] Schweitzer, P. A., *Counter-examples to the Seifert conjecture and opening closed leaves of foliations,* Ann. Math (2) 100(1974), pp. 386-400.

[31] Seifert, G., *Stability conditions for the existence of almost periodic solutions of almost periodic systems,* J. Math. Anal. Appl. 10(1965), pp. 409-418.

[32] Sell, G. R., *The structure of a flow in the vicinity of an almost periodic motion,* J. Differential Eqs. 27(1978) pp. 359-393.

[33] Smale, S., *Differentiable dynamical systems*, Bull. Amer. Math. Soc., 73(1967), pp. 747-817.

[34] Veech, W., *Point distal flows*, Amer. J. Math. 92(1970), pp. 205-242.

[35] Williams, R. F., *One-dimensional nonwandering sets*, Topology 6(1967), pp. 473-487.

[36] Yoshizawa, T., *Asymptotically almost periodic solutions of an almost periodic system*, Funk. Ekvac. 12(1969), pp. 23-40.

[37] Zaidman, S., *Soluzioni quasi-periodiche per alcune equazioni differenziali in spazi hilbertiani*, Ricerche Mat. 13(1964), pp. 118-134.

ON EXISTENCE OF PERIODIC SOLUTIONS FOR NONLINEARLY PERTURBED CONSERVATIVE SYSTEMS

Shair Ahmad

Oklahoma State University

Jorge Salazar

Colegio Universitario de Caracas

INTRODUCTION

Consider the second order vector differential equation

$$x'' + \text{grad } G(x) = p(t,x), \tag{1}$$

where $G \in C^2(\mathbb{R}^n, \mathbb{R})$, and $p \in C(\mathbb{R} \times \mathbb{R}^n, \mathbb{R}^n)$ is 2π-periodic in t for each fixed $x \in \mathbb{R}^n$.

W. S. Loud [6] considered the scalar version $x'' + g(x) = E \cos t$ of (1), and proved that if $g(x)$ is an odd function of class C^1, and if

$$(N+\delta)^2 \le g'(x) \le (N+1-\delta)^2 ,$$

with N an integer and $\delta > 0$, then the scalar version has a unique 2π-periodic solution. D. E. Leach [4] partially generalized Loud's theorem, by using the same inequality. A. C. Lazer and D. A. Sánchez [2] proved the existence of a 2π-periodic solution of the system

$$x'' + \text{grad } G(x) = p(t) = p(t+2\pi), \tag{2}$$

under the condition that the Hessian matrix $H(x)$ satisfy the inequality $sI \le H(x) \le qI$ for every $x \in \mathbb{R}^n$, where q,s are numbers such that $N^2 < s \le q < (N+1)^2$. A. C. Lazer [3] and

ISBN 0-12-045550-1

S. Ahmad [1] proved the uniqueness and existence, respectively, of a 2π-periodic solution of the equation (2) under less restrictive conditions than used by Lazer and Sánchez. J. R. Ward [8] proved the existence of a 2π-periodic solution of equation (1), weakening their inequality condition on $H(x)$ by requiring it to hold only outside a ball and requiring a Lipschitz condition on $p(t,x)$. Here, we shall establish this result without the assumption that $p(t,x)$ be Lipschitz.

Ward's result is the following:

Theorem 1. Let $G \in C^2(\mathbb{R}^n, \mathbb{R})$ and $p \in C(\mathbb{R} \times \mathbb{R}^n, \mathbb{R}^n)$, with $p(t+2\pi,x) = p(t,x)$ for all $(t,x) \in \mathbb{R} \times \mathbb{R}^n$. Assume that

(i) $\lim_{|x| \to \infty} \frac{|p(t,x)|}{|x|} = 0$ uniformly in t.

(ii) There exist constant symmetric $n \times n$ matrices A and B such that if $\lambda_1 \le \lambda_2 \le \dots \le \lambda_n$ and $\mu_1 \le \mu_2 \le \dots \le \mu_n$ are the eigenvalues of A and B, respectively, then there exist integers $N_k \ge 0$ $(k = 1,2,\dots,n)$ satisfying
$$N_k^2 < \lambda_k \le \mu_k < (N_k+1)^2.$$

(iii) There exists a number $M > 0$ such that $|p(t,x) - p(t,y)| \le M|x-y|$ for all $t \in \mathbb{R}$ and $x,y \in \mathbb{R}^n$.

(iv) There is a number $r > 0$ such that for all $a \in \mathbb{R}^n$ with $|a| \ge r$, the inequality
$$A \le \left(\frac{\partial^2 G(a)}{\partial x_i \partial x_j}\right) \le B$$ is satisfied.

Then equation (1) has a 2π-periodic solution.

First, we state two theorems which are basic and necessary for the development of our result. Since these results are known, we shall omit their proofs and simply indicate appropriate references.

Theorem 2 [3, p.91]. Let $Q(t)$ be a real $n \times n$ symmetric matrix-valued function with elements defined, measurable and 2π-periodic on the real line. Suppose there exist real, constant, symmetric matrices A and B such that $A \le Q(t) \le B$, $t \in (-\infty, \infty)$, and satisfying assumption (ii) of Theorem 1. Then there is no nontrivial 2π-periodic solution of the vector differential equation $w''(t) + Q(t)w(t) = 0$, where we understand a solution to be a function w with w' absolutely continuous such that w satisfies the differential equation a.e.

Actually in [3], Theorem 2 was established for $Q(t)$ continuous but a very slight modification of the proof in [3] gives the stronger result.

Theorem 3 [7, p. 27]. Let $\mathcal{B}$ be a normed space, B the closed unit ball in $\mathcal{B}$. Let T be a continuous compact mapping of B into $\mathcal{B}$ such that $T(\partial B) \subset B$. Then T has a fixed point.

Specifically, the following theorem will be proved:

Theorem 4. Consider the differential system (1), where $G \in C^2(\mathbb{R}^n, \mathbb{R})$, and $p \in C(\mathbb{R} \times \mathbb{R}^n, \mathbb{R}^n)$ is a 2π-periodic in t for each fixed $x \in \mathbb{R}^n$. Assume that the conditions (i), (ii) and (iv) of Theorem 1 hold. Then (1) has a 2π-periodic solution.

As a preliminary step in proving Theorem 4, we shall first establish the following:

Proposition 5. If for each $x \in \mathbb{R}^n$, $S(x)$ is an $n \times n$ symmetric matrix such that $A \leq S(x) \leq B$, where A and B are symmetric matrices satisfying assumption (ii) in Theorem 1, if the elements of S are continuous functions of x, $x \in \mathbb{R}^n$, and if $q : \mathbb{R} \times \mathbb{R}^n \to \mathbb{R}^n$ is a continuous function such that $q(t + 2\pi, x) = q(t,x)$ for all $(t,x) \in \mathbb{R} \times \mathbb{R}^n$ with $\frac{|q(t,x)|}{|x|} \to 0$ as $|x| \to \infty$ uniformly in t, then there exists a 2π-periodic solution of the differential equation

$$x'' + S(x)x = q(t,x). \tag{3}$$

Before discussing the proof, it might be more illuminating to present the basic reasoning upon which it depends in abstract form, first.

To prove the proposition we consider for each continuous 2π-periodic $\mathbb{R}^n$-valued function $u(t)$ the linear differential equation $x'' + S(u(t))x = q(t,u(t))$. Since, by Theorem 2, there is a unique 2π-periodic solution, this defines a mapping of a certain Banach space into itself, which is compact and continuous. Theorem 3 is used to guarantee a fixed point of this mapping, which yields a 2π-periodic solution of (3). Theorem 4 is then proved by showing that (1) can be written in the form (3).

Proof of Proposition 5. Consider the Banach space $(C_{2\pi}, \|\cdot\|)$ consisting of all continuous $\mathbb{R}^n$-valued 2π-periodic functions with the sup norm. For fixed $u \in C_{2\pi}$ the linear differential equation $x'' + S(u(t))x = q(t,u(t))$ has a unique 2π-periodic solution which we denote by Tu. This follows from the fact that, according to Theorem 2, the corresponding homogeneous differential equation has only the trivial 2π-periodic solution. This defines an operator $T : C_{2\pi} \to C_{2\pi}$. We shall prove that T is a compact operator, i. e., for each bounded sequence $\{u_m\}$ in $C_{2\pi}$, the sequence $\{Tu_m\} = \{v_m\}$ contains a subsequence converging to some limit in $C_{2\pi}$. We do this in two steps: (a) proving that $\{v_m\}$ is bounded, (b) proving that $\{v_m\}$ is an equicontinuous family of functions, so that we can apply Ascoli's Lemma. Suppose that $\{v_m\}$ is not bounded. By replacing the sequence $\{v_m\}$ by a suitable subsequence, we may assume $w_m^2 = \max_{t \in [0,2\pi]} \{|v_m(t)|^2 + |v_m'(t)|^2\} \to \infty$ as $m \to \infty$. Here $w_m > 0$. If we let $z_m = \frac{1}{w_m} v_m$ and $q_m(t) = \frac{1}{w_m} q(t, u_m(t))$, then we have

$$z_m'' + S(u_m(t)) z_m(t) = q_m(t), \quad \max_{t \in [0,2\pi]} \{|z_m(t)|^2 + |z_m'(t)|^2\} = 1.$$

Clearly, the family of functions $\{z_m\}$ is uniformly bounded and equicontinuous, also the family $\{z_m'\}$ satisfies the same properties. Therefore, there exists a subsequence $\{z_{m_k}\}$ of $\{z_m\}$ and functions $z, w \in C_{2\pi}$ such that $z_{m_k} \to z$ and $z_{m_k}' \to w$ uniformly. Obviously, $w = z'$. Since $z_m(t) = z_m(0) + \int_0^t z_m'(s)ds$, then $z(t) = z(0) + \int_0^t w(s)ds$. Set $S(u_m(t)) = (s_{ijm}(t))$. Since the elements of $S(u_m(t))$ are bounded, we may assume, without loss of generality that $s_{ijm_k}(t)$ converges weakly to $s_{ij}(t)$ in $L^2[0,2\pi]$ for $1 \le i, j \le n$. We want to show if $S(t) = (s_{ij}(t))$, then $A \le S(t) \le B$. With each symmetric matrix $\Sigma = (\sigma_{ij})$ we associate the point $(\sigma_{11}, \sigma_{12}, \ldots, \sigma_{1n}, \sigma_{22}, \ldots, \sigma_{2n}, \ldots, \sigma_{nn})$ in R^p with $p = \frac{n(n+1)}{2}$. With this identification the set $H = \{\Sigma : \Sigma$ is a symmetric matrix and $A \le \Sigma \le B\}$ is a compact convex subset of R^p. In view of Lemma 2.1 [5, p. 157] it follows that $S(t)$ is a symmetric matrix and satisfies $A \leq S(t) \leq B$. Note that $z \not\equiv 0$, because the condition

$\max_{t \in [0,2\pi]} \{|z_{m_k}(t)|^2 + |z'_{m_k}(t)|^2\} = 1$ and the uniform convergence imply that $\max_{t \in [0,2\pi]} \{|z(t)|^2 + |z'(t)|^2\} = 1$. For each k, $k = 1,2,\ldots$, the equation $z''_{m_k} + S(u_{m_k}(t))z_{m_k} = q_m(t)$ gives

$$z'_{m_k}(t) = z'_{m_k}(0) - \int_0^t S(u_{m_k}(s))z_{m_k}(s)ds + \int_0^t q_{m_k}(s)ds. \quad (4)$$

Since $q_{m_k}(t) = \dfrac{q(t,u_{m_k}(t))}{w_{m_k}}$, the boundedness of the sequence $\{u_{m_k}\}$ implies that $q_{m_k}(t) \to 0$ as $k \to \infty$. Moreover the weak convergence of the elements of $S(u_{m_k}(t))$ to the corresponding elements of $S(t)$ and the uniform convergence of $z_{m_k}(t)$ to $z(t)$ imply that

$$\int_0^t S(u_{m_k}(s))z_{m_k}(s)ds$$

$$= \int_0^t S(u_{m_k}(s))[z_{m_k}(s) - z(s)]ds$$

$$+ \int_0^t S(u_{m_k}(s))z(s)ds \to \int_0^t S(s)z(s)ds \quad \text{as } k \to \infty.$$

Therefore, letting $k \to \infty$ in (4), we obtain for each $t \in [0,2\pi]$, $z'(t) = z'(0) - \int_0^t S(s)z(s)ds$. Hence $z''(t) + S(t)z(t) = 0$ almost everywhere and, since $z \not\equiv 0$ and z is 2π-periodic, this contradicts Theorem 2. This contradiction proves that

T is compact. Continuity of T follows from standard continuity theorems in the theory of ordinary differential equations.

Now, we are going to show that $\frac{\|Tu\|}{\|u\|} \to 0$ as $\|u\| \to \infty$. Assuming the contrary, there exists a sequence $\{u_m\}$ in $C_{2\pi}$ and a number $c > 0$ such that $\|u_m\| \to \infty$ and $\|Tu_m\| > c\|u_m\|$. Writing $v_m = Tu_m$ and $z_m = \frac{1}{\|v_m\|}v_m$, we have

$$z_m''(t) + S(u_m(t))z_m(t) = \frac{1}{\|v_m\|} q(t,u_m(t)). \tag{5}$$

We claim that

$$\frac{1}{\|v_m\|} q(t,u_m(t)) \to 0 \quad \text{as} \quad m \to \infty \tag{6}$$

uniformly with respect to t, $t \in [0,2\pi]$. To see this, let $\varepsilon > 0$ be arbitrary. Since $\lim_{|x|\to\infty} \frac{|q(t,x)|}{|x|} = 0$ uniformly in t, there exists $L > 0$ such that $\frac{|q(t,x)|}{|x|} < \frac{\varepsilon}{2}$ if $|x| \geq L$. Consequently, if $k(\varepsilon) > 0$ is the maximum of $|q(t,x)| - \frac{\varepsilon}{2}|x|$ for $|x| \leq L$ then $|q(t,x)| \leq \frac{\varepsilon}{2}|x| + k(\varepsilon)$ for all x and t, in particular, $|q(t,u_m(t))| \leq \frac{\varepsilon}{2}|u_m(t)| + k(\varepsilon)$ for all t and m. Therefore, since $|u_m(t)| \leq \|u_m\|$ and $\|u_m\| \to \infty$ as $m \to \infty$, we see that $\frac{|q(t,u_m(t))|}{\|u_m\|} \leq \frac{\varepsilon}{2} + \frac{k(\varepsilon)}{\|u_m\|} < \varepsilon$ for m sufficiently large, which implies that

$$\frac{|q(t,u_m(t))|}{\|v_m\|} = \frac{|q(t,u_m(t))|}{\|u_m\|} \cdot \frac{\|u_m\|}{\|v_m\|} < \frac{\varepsilon}{c} \quad \text{for sufficiently}$$

large m. This proves (6). Since $\|z_m\| = 1$, it follows from (5) and the condition $A \leq S(u_m(t)) \leq B$ that the sequence

$\{z''_m(t)\}$ is uniformly bounded. This implies boundedness of the sequence $\{z'_m(t)\}$ which in turn implies that both of the sequences $\{z_m(t)\}$ and $\{z'_m(t)\}$ are equicontinuous and uniformly bounded. Applying our previous reasoning, we infer the existence of a sequence of integers $\{m_k\}$, a measurable 2π-periodic matrix function $S(t)$, and C^1 function $w \in C_{2\pi}$, with w' absolutely continuous such that $A \le S(t) \le B$ a.e., $\lim_{k\to\infty} z_{m_k}(t) = w(t)$, $\lim_{k\to\infty} z'_{m_k}(t) = w'(t)$ uniformly on $[0,2\pi]$, and $w''(t) + S(t)w(t) = 0$ a.e. But since $\|w\| = \lim_{k\to\infty} \|z_{m_k}(t)\| = 1$, this contradicts Theorem 2. This contradiction shows that $\frac{\|Tu\|}{\|u\|} \to 0$ as $\|u\| \to \infty$. Assume R to be so large that $\frac{\|Tu\|}{\|u\|} < 1$ if $\|u\| = R$. If B_R is the closed ball in $C_{2\pi}$ of radius R, then the compact, continuous operator T maps the boundary of B_R into B_R. By Theorem 3, there exists $u \in B_R \subset C_{2\pi}$ such that $Tu = u$. So the proof is complete.

<u>Proof of Theorem</u> 4. Let $H(x) = \left(\frac{\partial^2 G(x)}{\partial x_i \partial x_j}\right)$ be the Hessian matrix. By replacing $p(t,x)$ by $p(t,x) - \operatorname{grad} G(0)$, we may assume without loss of generality that $\operatorname{grad} G(0) = 0$, and hence that (1) has the form

$$x'' + M(x)x = p(t,x) \tag{7}$$

where $M(x) = \int_0^1 H(sx)ds$. We will show that there exist a number $R > 0$ and matrices $\tilde{A}$ and $\tilde{B}$ such that for all $x \in \mathbb{R}^n$, with $|x| \ge R$,

$$\tilde{A} \le M(x) \le \tilde{B} \tag{8}$$

and

$$N_k^2 < \lambda_k(\tilde{A}) \leq \lambda_k(\tilde{B}) < (N_k + 1)^2, \tag{9}$$

where the integers N_k are the same as above ($\lambda_k(C)$ denotes the kth eigenvalue of the matrix C). Let $\varepsilon > 0$ be any number in the interval $(0,1)$. Let $L = \max_{|y| \leq r} |H(y)|$.

If $|x| \geq \frac{r}{\varepsilon}$ and $s \geq \varepsilon$ then $A \leq H(sx) \leq B$, because $|sx| = |s||x| \geq \varepsilon \frac{r}{\varepsilon} = r$. Let $|x| \geq \frac{r}{\varepsilon}$ and $v \in \mathbb{R}^n$; then

$$\begin{aligned}
<v,M(x)v> &= \int_0^1 < v,H(sx)v > ds \\
&= \int_0^\varepsilon < v,H(sx)v > dx + \int_\varepsilon^1 < v,H(sx)v > ds \\
&\leq \int_0^\varepsilon < v,Lv > ds \quad \int_\varepsilon^1 < v,Bv > ds \\
&= \varepsilon < v,Lv > + (1-\varepsilon) < v, Bv > \\
&= < v ,(\varepsilon LI + (1-\varepsilon)B)v > .
\end{aligned}$$

Similarly, if $|x| \geq \frac{r}{\varepsilon}$ and $v \in \mathbb{R}^n$,

$$\begin{aligned}
<v,M(x)v> &= \int_0^1 < v,H(sx)v > ds \\
&= \int_1^\varepsilon < v,H(sx)v > ds + \int_\varepsilon^1 < v,H(sx)v > ds \\
&\geq -\int_0^\varepsilon < v,Lv > ds + \int_\varepsilon^1 < v,Av > ds \\
&= < v,(\varepsilon LI + (1-\varepsilon)A)v > .
\end{aligned}$$

This shows that $|x| \geq \frac{r}{\varepsilon}$ implies

$(I-\varepsilon)A - \varepsilon LI \leq M(x) \leq (1-\varepsilon)B + \varepsilon LI$. If $\tilde{A} = (1-\varepsilon)A - \varepsilon LI$ and $\tilde{B} = (1-\varepsilon)B + \varepsilon LI$ then

$$\lambda_j(\tilde{A}) = (1 - \varepsilon)\lambda_j(A) - \varepsilon L,$$

$$\lambda_j(\tilde{B}) = (1 - \varepsilon)\lambda_j(B) + \varepsilon L,$$

$j = 1,2,\ldots,n$, and thus (8) and (9) hold if $\varepsilon > 0$ is sufficiently small. Choose such a number ε and set $R = \frac{r}{\varepsilon}$. Then we have reduced the problem to a consideration of (7), (8) and (9) for $|x| \geq R$. Now let $\phi : [0,\infty) \to \mathbb{R}$ be a real-valued function defined by

$$\phi(s) = \begin{cases} 1, & 0 \leq S \leq R \\ 2 - \frac{S}{R}, & R < S \leq 2R \\ 0, & S > 2R. \end{cases}$$

Then (7) can be written as

$$x'' + S(x)x = q(t,x),$$

where $S(x) = [1 - \phi(|x|)]M(x) + \phi(|x|)\tilde{A}$ and $q(t,x) = p(t,x) + \phi(|x|)[\tilde{A} - M(x)]x$. The following conditions are now satisfied:

a) $S(x)$ is symmetric.

b) $\tilde{A} \leq S(x) \leq \tilde{B}$ for all $x \in \mathbb{R}^n$. This follows since if $|x| \leq R$, $S(x) = \tilde{A}$, and if $|x| \geq 2R$, $S(x) = M(x)$. Furthermore, if $R \leq |x| < 2R$, we have $0 \leq \phi(|x|) \leq 1$. So, for any $v \in \mathbb{R}^n$,

$$\begin{aligned} \langle v,S(x)v \rangle &= \langle v,[1 - \phi(|x|)]M(x)v \rangle + \langle v,\phi(|x|)\tilde{A}v \rangle \\ &\geq \langle v,[1 - \phi(|x|)\tilde{A}v \rangle + \langle v,\phi(|x|)\tilde{A}v \rangle \\ &= \langle v,\tilde{A}v \rangle \end{aligned}$$

Therefore $S(x) \geq \tilde{A}$. Similarly, $S(x) \leq \tilde{B}$.

c) $\lim_{|x| \to \infty} \frac{|q(t,x)|}{|x|} = 0$. For since

$$|q(t,x)| \leq |p(t,x)| + |\phi(|x|)||\tilde{A} - M(x)|.$$

we have

$$\frac{|q(t,x)|}{|x|} \leq \frac{|p(t,x)|}{|x|} + |\phi(|x|)||\tilde{A} - M(x)|.$$

By the hypothesis and the definition of ϕ, the right side goes to 0 as $|x| \to \infty$. Hence, by proposition 5, $x'' + S(x)x = q(t,x)$ has a 2π-periodic solution, and the proof is complete.

REFERENCES

[1] Ahmad, S., *An existence theorem for periodically perturbed conservative systems*, Mich. Math. J. 20 (1973), pp. 385-392.

[2] Lazer, A. C., and D. A. Sánchez, *On periodically perturbed conservative systems*, Mich. Math. J. 16 (1969), pp. 193-200.

[3] Lazer, A. C., *Application of a lemma on bilinear forms to a problem in nonlinear oscillations*, Proc. Amer. Math. Soc. 33 (1972), pp. 89-94.

[4] Leach, D. E., *On Poincaré's perturbation theorem and a theorem of W. S. Loud*, J. Differential Equations 7 (1970), pp. 34-53.

[5] Lee, E. B., and L. Markus, "Foundations of Optimal Control Theory," Wiley, New York, 1967.

[6] Loud, W. S., *Periodic solutions of nonlinear differential equations of Duffing type*, Proc. U. S.-Japan Seminar on Differential and Functional Equations (Minneapolis, Minn., 1967), pp. 199-224, Benjamin, New York, 1967.

[7] Smart, D. A., "Fixed Point Theorems," Cambridge University Press, New York, 1975.

[8] Ward, J. R., *The existence of periodic solutions for nonlinearly perburbed conservative systems*, Nonlinear Analysis 3 (1979), pp. 677-705.

START POINTS IN SEMI-FLOWS

Prem N. Bajaj

Wichita State University

INTRODUCTION

Semi-flows or semi-dynamical systems (s.d.s.) are defined only for future time. Natural examples of s.d.s. are provided by functional differential equations for which existence and uniqueness conditions hold. S.d.s. theory not only generalizes substantial part of Dynamical Systems theory, but also gives rise to many new and interesting notions, e.g. of a start point, singular point. The notion of a start point dates back at least 1953, [5]. In this paper we discuss some properties of start-points' sets. For a family of s.d.s., their product s.d.s. is defined in a natural way. It is possible that none of the factor s.d.s. to have a start point, but the product s.d.s. contains start points. This gives rise to the notion of an improper start point. We examine the conditions for the set of proper/improper start points to be everywhere dense. We also consider some of the connectedness properties of the sets of proper/improper start points. Finally, it is pointed out by means of examples, some of the implications that do not hold.

ISBN 0-12-045550-1

1. Definitions. A semi-dynamical system (s.d.s.) is a pair (X,π) where X is a topological space and π is a continuous map defined on $X \times R^+$ with values in X satisfying the identity axiom $\pi(x,0) = x$ and the semi-group property $\pi(\pi(x,t),s) = \pi(x,t + s)$ for all t,s in R^+ and x in X. Here R^+ denotes the set of nonnegative reals with usual topology. For brevity (e.g. as in [1]), $\pi(x,t)$ will be denoted by xt. Throughout, (X,π) will denote an s.d.s.

Define a map $E : X \to R^+ \cup \{+\infty\}$ by $E(x) = \sup \{t \geqq 0 : yt = x$ for some y in $X\}$. A point x in X is said to be a start point if $E(x)$, called the escape time of x, is zero.

A point x is (positive)critical if $xR^+ = \{x\}$. Positive trajectory $r^+(x)$ is defined to be the set $\{y : y = xt, t \in R^+\} = xR^+$.

2. Proposition. In a semi-dynamical system, the set S of start points has an empty interior. Consequently, if U is a non-empty open set, then $U - S$ is non-empty.

3. Proposition. Let (X_i, π_i), $i \in I$ be a family of semi-dynamical systems. Let $X = \prod_{i \in I} X_i$ be the product space. If $\pi : X \times R^+ \to X$ is defined by $\pi(x,t) = \{x_i t\}$, $x = \{x_i\}$, then (X,π) is a semi-dynamical system which we call product s.d.s.

Proof. Identity axiom and semi-group axiom are clear. To see that π is continuous, let x^j be a net in X, t^j a net in R^+ such that $x^j \to x$ and $t^j \to t$. Then $x_i^j \to x_i$ and so, $\pi(x^j, t^j) = \{x_i^j t^j\} \to \{x_i t\} = \pi(x,t)$.

4. Theorem. Let (X_i, π_i), $i \in I$ be a family of semi-dynamical systems and (X,π) their product s.d.s. If S_i is the set of start points in (X_i, π_i), $i \in I$ and S the corresponding set in (X,π), then $S \supset \bigcup_i (S_i \times \prod_{j \neq i} X_j)$. If I is finite then $S = \bigcup_i (S_i \times \prod_{j \neq i} X_j)$.

Proof: If x_i is a start point in (X_i, π_i) for some i, then $x \in S$, $x = \{x_i\}$ etc.

5. Remark. It is easily seen that x, $x = \{x_i\}$ can be a start point in (X,π) even though $x_i \notin S_i$ for each i. Indeed it is possible that none of the factor systems has any start point, but start points exist in the product s.d.s. This leads us to the

6. Definition. Let $(X_i \pi_i)$, $i \in I$ be a family of semi-dynamical systems and (X,π) their product s.d.s. Let $x \in X$, $x = \{x_i\}$ be a start point. Then relative to the factorization $X = \prod_{i \in I} X_i$, x is said to be a proper start point if x_i is a start point for some i in I: otherwise start point x is said to be improper.

7. Notation and Remarks. Throughout the rest of this paper, (X_i, π_i) $i \in I$ denotes a family of s.d.s. and (X,π) their product s.d.s. Moreover S_i, S denote, respectively, the sets of start points in (X_i, π_i) and (X,π). The set of proper start points in (X,π) is denoted by S^*; consequently $S - S^*$ denotes the set of improper start points.

It follows from Theorem 4 that improper start point can exist only if I is infinite. Further an x in X, $x = \{x_i\}$ is a start point if and only if both the following conditions hold.

(i) $x_i \notin S_i$ for each i in I

(ii) $\inf \{E(x_i) : i \in I\} = 0$.

Moreover, the condition (ii) can be replaced by the equivalent condition (ii)'. There exists a sequence $E(x_i)$ in $\{E(x_i) : i \in I\}$ converging to zero.

8. Theorem. In the product s.d.s. (X,π), there exists an improper start point if and only if for infinitely many i in I, the s.d.s. (X_i,π_i) contains a point x_i with finite non-zero escape time.

Proof. If $x \in X$, $x = \{x_i\}$ is an improper start point, then $E(x_i) > 0$ for every i. Clearly $\{i : E(x_i) < 1\}$ is infinite. Indeed, otherwise, $E(x) = \inf\{E(x_i) : i \in I\} > 0$; contradicts that x is a start point.

Conversely for each positive integer n, pick an i_n in I such that $0 < E(x_{i_n}) < +\infty$ for some x_{i_n}. Choose y_{i_n} in X_{i_n} such that $0 < E(y_{i_n}) \leqq \frac{1}{n}$. Let $T > 0$ be fixed. Then $z \in X$, $z = \{z_i\}$ such that $z_{i_n} = y_{i_n}$ for every n, and $z_i = x_i T$ otherwise where $x_i \in X_i$ is arbitrary, is an improper start point.

9. Theorem. The set of improper start points is either empty or dense in X.

Proof. Let $x = \{x_i\}$ be an improper start point. Let $U = U_{i_1} \times U_{i_2} \times \ldots \times U_{i_m} \times \prod_{i \neq i,..,i_m} X_i$ be a basic open set in open set in X. Pick $z \in X$, $z = \{z_i\}$ such that

$z_{i_j} \in U_{i_j} - S_{i_j}$, $j = 1, 2, \ldots, m$ and $z_i = x_i$ otherwise. It is easily seen that z is an improper start point. Since $z \in U$ and U is arbitrary, the result follows.

Considering the cases when the set of proper start points is dense in X, and combining with above theorem, we have

10. Theorem. The set of start points is dense in X if and only if at least one of the following holds:

(a) There exists an improper start point

(b) For some i in I, S_i is dense in X_i

(c) Infinitely many factor semi-dynamical systems contain start points.

Let us now consider some of the connectedness properties of the sets S^*, $S - S^*$, and S.

11. Theorem. Let $S_i \neq \not{c}$, $S_j \neq \not{c}$ for some i, j in I, $i \neq j$. Then the set of proper start points is (path) connected if and only if X is (path) connected.

12. Theorem. Let there exist an improper start point. Then the following are equivalent:

(a) X is (path) connected.

(b) The set of improper start points is (path) connected.

(c) The set of start points is (path) connected.

13. Remark. Even when improper start points exist, (path) connectedness of the set of proper start points neither implies nor is implied by the (path) connectedness of the product space. To see this we consider the following examples.

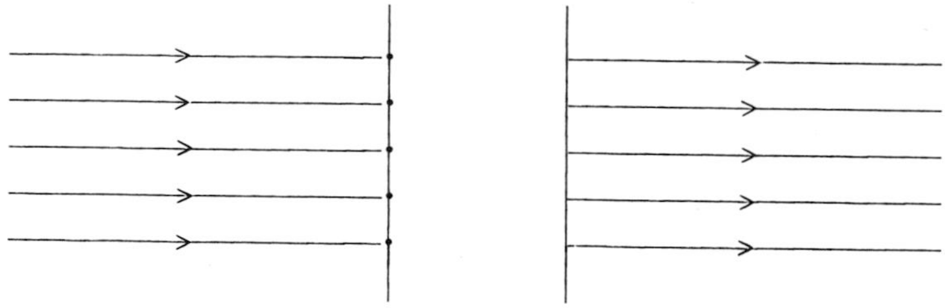

(X_1, π_1)

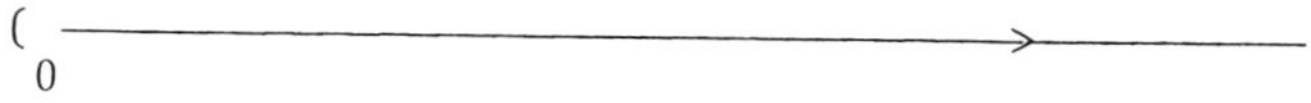

(X_i, π_i) , $i = 2, 3, 4, \ldots$.

Figure 1

14. <u>Example</u>. Let $X_1 = \{(x,y) : |x| \geqq 1\} \subset R^2$. On X_1 define a flow to the right, taking each point on the line $x = 1$ as a start point, and each point on the line $x = -1$ to be critical point. Let $X_i = (0, +\infty) \subset R$ for $i = 2,3,4,\ldots$, and define map π_i to be uniform flow to the right. (See Figure 1).

Let (X,π) be the product s.d.s. of the family (X_i,π_i), $i = 1, 2, \ldots$. Then the set of proper start point is path connected, but X is not.

15. <u>Example</u>. Let $X_1 = \{(x,y) : x \geqq 0$ or $|y| \leqq 1\} \subset R^2$. Define π_1 to be uniform flow to the right (taking the set $\{(0,y) : |y| > 1$ to be start point set. Let (X_i,π_i), $i = 2, 3, \ldots$ be as in above example. Let (X,π) be product s.d.s. of the family (X_i, π_i), $i = 1,2,3,\ldots$. (See Figure 2) . Clearly X is path connected, but S^* is not.

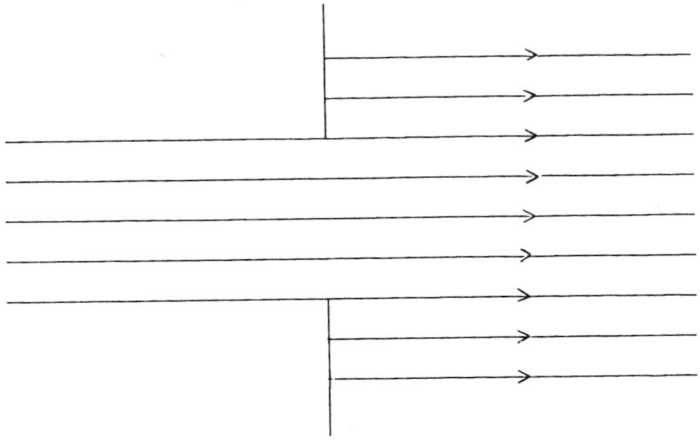

(X_1, π_1)

Figure 2

REFERENCES

[1] Auslander, Joseph, *Filter stability in dynamical systems*, SIAM J. Math. Anal., Vol. 8 (1977), pp. 573-579.

[2] Bajaj, Prem N., *Start points in semi-dynamical systems*, Funk. Ekv. 13 (1971), pp. 171-177.

[3] Bajaj, Prem N., *Connectedness properties of start points in semi-dynamical systems*, Funk. Ekv. 14 (1971), pp. 171-175.

[4] Bhatia, N. P. and O. Hajek, Local Semi-dynamical Systems, Springer-Verlag, New York, 1969.

[5] Lots, I. Flugge, Discontinuous Automatic Control, Princeton University Press 1953.

[6] Willard, S., General Topology, Addison-Wesley, Reading, Massachusetts, 1970.

A SADDLE-POINT THEOREM

Peter W. Bates
Texas A&M University

Ivar Ekeland
Université Paris-Dauphine

This note is to give a simple critical point theorem for certain indefinite functionals. The result follows from a theorem by Ekeland on minimization of nonconvex functionals [1] and provides a generalization of a recent minimax theorem due to Lazer, Landesman and Meyers [2].

Let H be a real Hilbert space and J a C^2 functional on H, with $J'(u)$ and $J''(u)$ denoting the first and second derivatives of J at $u \in H$.

Theorem 1. Suppose that there is some constant $m > 0$ such that for each $u \in H$, there exist two closed (not necessarily orthogonal) subspaces $H_1 = H_1(u)$ and $H_2 = H_2(u)$ with $H = H_1 \oplus H_2$ and

(i) $(J''(u)w, w) \geq m\|w\|^2$ for $w \in H_1$,

(ii) $(J''(u)w, w) \leq -m\|w\|^2$ for $w \in H_2$.

Then J has a critical point, i.e., there exists $u_0 \in H$ such that $J'(u_0) = 0$. In fact, J' maps onto H. Furthermore, if H_1 and H_2 are independent of u, then J is one-to-one.

ISBN 0-12-045550-1

The proof will be given below following two important results, one on approximate minimization and the other on surjectivity of mappings.

Theorem 2 (Ekeland). Let V be a complete metric space and $F : V \to \mathbb{R} \cup \{+\infty\}$ a lower semicontinuous function which is bounded below and not identically $+\infty$. Then for any $\varepsilon > 0$ there exists some point $v_\varepsilon \in V$ such that

$$F(v_\varepsilon) \le \inf F + \varepsilon \quad \text{and}$$

$$F(u) > F(v_\varepsilon) - \varepsilon d(v_\varepsilon, u) \quad \text{for all } u \in V.$$

For a concise proof see e.g. [1]. The following result is less well-known, so the proof is presented here. Suppose that X and Y are Banach spaces and that $\Phi : X \to Y$ is Gâteaux differentiable, with its derivative at $x \in X$ denoted by $\Phi'(x)$.

Theorem 3 (Ekeland. Suppose there exists a constant $k > 0$ such that for $x \in X$ and $y \in Y$ there exists $z \in X$ such that

$$\|z\| \le k\|y\| \quad \text{and} \quad \Phi'(x)z = y.$$

Then Φ is onto.

Proof. It suffices to show that 0 is in the range of Φ since $\Phi - y$ satisfies the hypotheses for each $y \in Y$. Consider the functional $F(x) \equiv \|\Phi(x)\|$ and apply Theorem 2 with $\varepsilon < 1/k$. There exists x_ε such that

$$\|\Phi(x_\varepsilon)\| \le \|\Phi(x)\| + \varepsilon\|x - x_\varepsilon\| \quad \text{for all } x \in X. \tag{1}$$

Let $v \in X$ and set $x = x_\varepsilon + tv$ for $t > 0$. Inequality (1) may be written

$$[\|\Phi(x_\varepsilon + tv)\| - \|\Phi(x_\varepsilon)\|]/t \geq -\varepsilon\|v\|. \tag{2}$$

Set $y_t = \Phi(x_\varepsilon + tv) \neq 0$ and let $y_t^* \in Y^*$ be such that $\|y_t^*\|_* = 1$ and $<y_t^*, y_t> = \|y_t\|$. Inequality (2) implies

$$<y_t^*, [\Phi(x_\varepsilon + tv) - \Phi(x_\varepsilon)]/t> \geq -\varepsilon\|v\|. \tag{3}$$

Since Φ is Gâteaux differentiable and the unit ball in Y^* is weak-star compact, allowing t to approach zero, there exists $y_0^* \in Y^*$ with $\|y_0^*\| \leq 1$ and

$$<y_0^*, \Phi(x_\varepsilon)> = \|\Phi(x_\varepsilon)\| \quad \text{and} \tag{4}$$

$$<y_0^*, \Phi'(x_\varepsilon)v> \geq -\varepsilon\|v\|. \tag{5}$$

Now by hypothesis, there exists $z \in X$ such that

$$\|z\| \leq k\|\Phi(x_\varepsilon)\| \quad \text{and} \quad \Phi'(x_\varepsilon)z = -\Phi(x_\varepsilon),$$

and so (4) and (5) imply

$$\|\Phi(x_\varepsilon)\| = <y_0^*, \Phi(x_\varepsilon)> \leq \varepsilon\|z\| \leq \varepsilon k\|\Phi(x_\varepsilon)\|.$$

Hence, $\Phi(x_\varepsilon) = 0$, which completes the proof.

Proof of Theorem 1. Consider the self-adjoint operator $J''(u) : H \to H$ for each $u \in H$. We show that $J''(u)$ is invertible with $\|J''(u)^{-1}\| \leq 2/m$. Then taking $\Phi = J'$, $X = Y = H$, $k = 2/m$ and $z = J''(x)^{-1}y$ in Theorem 3, the surjectivity of J' will follow. Let $J''(u)v = w$ with $v = v_1 + v_2$ where $v_1 \in H_1$ and $v_2 \in H_2$, and let $\bar{v} = v_1 - v_2$. Then because $J''(u)$ is self-adjoint $(J''(u)v, \bar{v}) = (w, \bar{v})$ gives

$$m(\|v_1\|^2 + \|v_2\|^2) \leq w\ (\|v_1\| + \|v_2\|) \quad \text{and}$$

$$\frac{m}{2}\|v\| \leq \frac{m}{2}(\|v_1\| + \|v_2\|) \leq \|w\|. \tag{6}$$

Thus, $J''(u)$ is one-to-one and self-adjoint so its range is dense in H. Since $J''(u)$ is bounded, inequality (6) implies that its range is closed and $J''(u)$ is invertible. Finally, suppose that the decomposition $H = H_1 \oplus H_2$ is independent of u, and that $J'(u) = J'(w)$ for some $u, w \in H$. Write $u = u_1 + u_2$, $w = w_1 + w_2$ with $u_1, w_1 \in H_1$ and $u_2, w_2 \in H_2$. Let $\bar{u} = u_1 - u_2$ and $\bar{w} = w_1 - w_2$, then

$$0 = [J'(w) - J'(u)](\bar{u} - \bar{w})$$

$$= \int_0^1 < J''(u + s(w - u))(w - u), (\bar{u} - \bar{w}) > ds$$

$$\leq -m(\|w_1 - u_1\|^2 + \|w_2 - u_2\|^2).$$

Consequently, $u = w$ and J' is one-to-one.

Remarks. The advantages of Theorem 1 over the main result of [2] are that neither H_1 nor H_2 is required to be finite dimensional and that the decomposition of H as $H_1 \oplus H_2$ may vary from point to point and the angle between the subspaces is not necessarily bounded away from zero. Perhaps a disadvantage with Theorem 1 is that the critical point is not characterized explicitly as a minimax as in [2].

REFERENCES

[1] Ekeland, Ivar, *Nonconvex minimization problems*, Bull. (New Series) AMS 1 (1979), pp. 443-474.

[2] Lazer, A. C., E. M. Landesman, and D. R. Meyers, *On saddle point problems in the calculus of variations, the Ritz algorithm, and monotone convergence*, J. Math. Anal. Appl. 52 (1975), pp. 594-614.

GENERALIZED HOPF BIFURCATION IN R^n AND h-SYMPTOTIC STABILITY

S. R. Bernfeld[1]

University of Texas at Arlington

L. Salvadori

Università di Trento, Italy

INTRODUCTION

The concept and analysis of structural stability of two dimensional autonomous differential systems was provided by Andronov and Pontryagin in a fundamental paper [1] in 1937. Within the next fifteen years Andronov and his group (including Leontovich, Chaikin, Witt, et al.) developed a major part of the present theory of two dimensional structurally stable and unstable system of differential equations.

Motivated by problems in classical and celestial mechancis, mathematicians and physicists have been very concerned with providing a qualitative description of the trajectories of perturbed systems in a neighborhood of a structurally unstable equilibrium point in $R^n (n \geq 2)$.

To be more specific let us consider a differential system

$$\dot{x} = f_o(x), \tag{1}$$

[1]Research partially supported by C.N.R. (National Council of Research in Italy), by U.S. Army Grant DAAG-29-77-G0062 and by University of Texas at Arlington Organized Research Grant.

ISBN 0-12-045550-1

where $f_0 \in C^\infty[B^n(r_0),R^n]$, $f_0(0) = 0$, $B^n(r_0) = \{x \in R^n : \|x\| < r_0\}$. Assume that the Jacobian matrix $f_0'(0)$ has two purely imaginary eigenvalues $\pm i$ and that the remaining eigenvalues $\{\lambda_j\}_{j=1}^{n-2}$ satisfy $\lambda_j \neq mi$ $(m = 0, \pm 1, \pm 2, \ldots)$. For those $f \in C^\infty[B^n(r_0), R^n]$, $f(0) = 0$, which are close to f_0 (in an appropriate topology) consider the pertrubed system

$$\dot{x} = f(x). \tag{2}$$

We shall now address ourselves to the following local problem.

<u>Problem</u>. For those f near f_0 give a complete description of the compact invariant sets of (2) lying in $B^n(r)$ for r small.

In approaching this problem we will consider for any positive integer k the following two properties:

(a) there exists a neighborhood N of f_0 and two numbers $r_1 > 0$, and $\delta_1 > 0$ such that for every $f \in N$ there are at most k nontrivial periodic orbits of (2) lying in $B^n(r_1)$ having periods in $[2\pi - \delta_1, 2\pi + \delta_1]$;

(b) for each integer j, $0 \leq j \leq k$, for each $r \in (0,r_1]$ and $\delta \in (0, \delta_1]$ and for each neighborhood N of f_0, $N \subset N$, there exists an $f \in N$ such that (2) has exactly j nontrivial periodic orbits lying in $B^n(r)$ and having period in $[2\pi - \delta, 2\pi + \delta]$.

In R^2, Andronov et al. [2] provided a partial answer to the Problem by proving that properties (a) and (b) are a consequence of the origin of (1) being $(2k + 1)$ - asymptotically stable or $(2k + 1)$-completely unstable. The origin of system (1) is said to be $(2k+1)$-asymptotically stable ($(2k+1)$-completely unstable) if $2k+1$ is the smallest positive integer such that the origin of (2) is asymptotically stable (completely unstable) for all f for which $f(x) - f_0(x)$ is $0(|x|^{2k+2})$ (that is, the stability property is recognizable by terms up to order $2k+1$ in the Taylor expansion of f_0) (see [10] for further information on $(2k+1)$ - asymptotic stability). There is an equivalence between the $(2k+1)$ - asymptotic stability ($(2k+1)$ - complete instability) of the origin of (1) and the existence of a computable Liapunov function $V(x)$ whose derivative along solutions of (1), denoted by $\dot{V}(x)$, is negative definite and satisfies $\dot{V}(x) = G_{2k+2}|x|^{2k+2} + 0(|x|^{2k+3})$ where G_{2k+2} is negative (positive) computable constant (see [10]). This algebraic procedure of computing $V(x)$ is originally due to Poincaré [12]. In [10] Negrini and Salvadori related the $(2k+1)$ - asymptotic stability ($(2k+1)$ - complete instability) of the origin of (1) with the existence of asymptotically stable (completely unstable) bifurcating period orbits of (2) when the class of perturbations f is restricted to a one parameter family $\{f_\mu\}$, $0 \le \mu < 1$ (the usual Hopf bifurcation) such that the origin of (2) is completely unstable (asymptotically stable) for $\mu > 0$ and this stability property is recognizable by the linear part of f_μ. They showed for each $\mu > 0$ (μ small) there exists exactly one periodic orbit of $\dot{x} = f_\mu(x)$ with period near 2π and lying close to the origin.

In recent works [3], [4], the authors extend the results of Andronov et al. by proving properties (a) and (b) for

$n = 2$ are equivalent to the $(2k + 1)$ - asymptotic stability ($(2k + 1)$ - complete instability) of the origin of (1). We also prove for $n = 2$ that the $(2k + 1)$ - asymptotic stability of the origin of (1) ($(2k + 1)$ - complete instability) implies

(c) $$N = \bigcup_{j=0}^{k} S_j,$$

and each S_j has a nonempty interior with f_o lying on its boundary for each j, where $S_j = \{f \in N$ such that (2) has exactly j nontrivial periodic orbits lying in $B^n(r_1)$ with period in $[2\pi - \delta_1, 2\pi + \delta_1]\}$;

(d) as $f \to f_o$ the set of periodic orbits lying in $B^n(r_2)$ with period in $[2\pi - \delta_1, 2\pi + \delta_1]$ tend to $\{0\}$ in the Hausdorff topology.

In the generic case, $k = 1$, necessary and sufficient conditions were given for those f contained in S_0 and in S_1 in terms of the Poincaré constant G_4 and the real part of the eigenvalues of the Jacobian $f'(0)$ (which we refer to as α). Namely $f \in S_1$ if and only if $\alpha G_4 < 0$ ($f \in S_0$ if and only if $\alpha G_4 \geq 0$).

Finally we consider the transcendental case when $G_i = 0$ for all i; that is the origin of (1) is not $(2k + 1)$ - asymptotically stable nor $(2k + 1)$ - completely unstable for any positive integer k. We provide a characterization of the trajectories of (2) in a suitable neighborhood of the origin. We omit the details here but point out that if the functions f_o is analytic (as assumed by Andronov et al. [2]) then $G_i = 0$ for all i is equivalent to the origin of (1) being a center. This is not true in the C^∞ case and a rather extensive treatment of this case can also be found in [5].

In 1978, prior to our papers [3] and [4], Chafee [6] considered in R^n the same aspect of the Problem as Andronov et al. Using the alternative method [7], he obtained a determining equation $\psi(\xi, f) = 0$, where ξ is a measure of the amplitude of the bifurcating orbit of (2) and f represents the right-hand side of (2). He found that $\psi(0,f_0) = 0$ and by assuming that the multiplicity of the zero root of $\psi(\cdot, f_0) = 0$ is equal to some number k he proved that conclusions (a), (b), (c) and (d) hold. Since the determination of $\psi(\cdot,\cdot)$ is only known implicity the determination of k in Chafee's results is a new problem in itself. Thus for the case $n = 2$ we have shown [4] that the k of Chafee [6] is precisely equal to the k associated with the $(2k + 1)$ - asymptotic stability or $(2k + 1)$ - complete instability of the origin of (1).

We are now attempting to characterize the k of Chafee in R^n, $n \geq 3$. First suppose that $f_0'(0)$ has two purely imaginary eigenvalues $\pm i$ and the remaining eigenvalues have nonzero real parts. By using the Center Manifold Theorem [9], we than recognize that there exists a 2-dimensional manifold, say H_0, tangent at the origin 0 of R^n to the eigenspace corresponding to the eigenvalues $\pm i$, which is locally invariant with respect to the unperturbed equation (1). In this case we have the following characterization of properties (a) and (b).

<u>Theorem 1</u>. A necessary and sufficient condition for properties (a) and (b) to occur is that 0 is for equation (1)

either (2k + 1) - asymptotically stable or (2k + 1) - completely unstable on H_0 (i.e. with respect to initial points lying on H_0).

Under some more particular hypotheses on the spectrum of $f_o'(0)$, the stability properties in Theorem 1 can be expressed directly in terms of the unperturbed equation (1), without any explicit involvement of H_0. Precisely the following theorem holds.

Theorem 2. Suppose that $f_o'(0)$ has two purely imaginary eigenvalues $\pm i$ and the remaining eigenvalues have negative real parts. Then a necessary and sufficient condition for properties (a) and (b) to occur is that 0 is for equation (1) either (2k + 1) - asymptotically stable or (2k + 1) - unstable (in the whole).

(Notice that we are using the concept of (2k + 1) - instability whose definition is analogous to that of (2k + 1) - complete instability). A similar theorem can be stated when $f_o'(0)$ has two purely imaginary eigenvalues $\pm i$ and the remaining eigenvalues have positive real parts. Our proof of Theorem 2 involve the extension of the Poincaré procedure given by Liapounov in his treatise on stability of motion [8]- in order to prove the equivalence between the stability properties of 0 for equation (1) occuring in Theorem 1 and in Theorem 2, respectively.

The much more difficult problem in R^n concerns the case in which $f_o'(0)$ has 2p purely imaginary eignevalues ($4 \leq 2p \leq n$). In this situation the Center Manifold is 2p dimensional and the Poincare-Bendixson theory is no longer applicable. In this case we are using a generalization in R^n of the Poincaré procedure given in [11]. Some results have been obtained for the existence of quasiperiodic orbits as well as for periodic orbits under a suitable non-resonance condition on the eigenvalues (see Liapounov [8]) (this work as well as Theorems 1 and 2 is being done jointly with P. Negrini).

We finally note that there still remain many interesting and important questions concerning the Problem.

REFERENCES

[1] Andronov A., and L. Pontryagin, *Systèmes grossiers*, Dokl. Akad. Nauk. SSR, 14(1937), pp. 247-251.

[2] Andronov, A., E. Leontovich, I. Gordon, and A. Maier, "Theory of Bifurcations of Dynamical Systems in the Plane," Israel Program of Scientific Translations, 1971, Jerusalem.

[3] Bernfeld, S., and L. Salvadori, *Generalized Hopf bifurcation*, Conference on Recent Advances in Differential Equations, Trieste, Italy (to appear).

[4] Bernfeld, S., and L. Salvadori, *Generalized Hopf bifurcation and h-asymptotic stability*, J. of Nonl. Anal. T.M.A. (to appear).

[5] Bernfeld, S., and L. Salvadori, *Generic properties of h-asymptotically stable systems and Hopf bifurcation*, (to appear).

[6] Chafee, N., *Generalized Hopf bifurcation and perturbation in a full neighborhood of a given vector field*, Indiana, Univ. Math. J., 27(1978), pp. 173-194.

[7] Hale, J., "Ordinary Differential Equations," Interscience, New York, 1969.

[8] Liapounov, A. M., "Probléme general de stabilite du mouvement," Ann. of Math. Studies, 17, Princeton University Press, Princeton, New Jersey, 1947.

[9] Marsden, G., and M. McCracken, "The Hopf Bifurcation and its Applications," Notes in Applied Math. Sci. 19, Springer, New York, 1976.

[10] Negrini, P., and L. Salvadori, "Attractivity and Hopf Bifurcation," J. Nonlinear Anal., TMA, 3(1979), pp. 87-100.

[11] Salvadori, L., *Sulla stabilità dell'equilibrio nei casi critici*, Ann. Mat. Pura Appl. (4) LXIL(1965), pp. 1-33.

[12] Sansone, G., and R. Conti, "Nonlinear Differential Equations," Macmillan New York, 1964.

THE POINCARÉ-BIRKHOFF "TWIST" THEOREM AND PERIODIC SOLUTIONS OF SECOND-ORDER NONLINEAR DIFFERENTIAL EQUATIONS

G. J. Butler

University of Alberta

INTRODUCTION

Let $f(t,x)$ be a continuous real-valued function on $\mathbb{R}^2$, periodic in t with period $\omega > 0$, have the same sign as x for $|x|$ sufficiently large (independently of t) and satisfy $f(t,0) \equiv 0$. Suppose that f is sufficiently smooth for the uniqueness of solutions of initial value problems of the differential equation

$$x''(t) + f(t,x(t)) = 0 . \tag{1}$$

We shall refer to all of the above assumptions on f as condition (a).

Consider the problem of the existence of nontrivial periodic solutions of (1). If the equation is linear, that is, $f(t,x(t)) = p(t)x(t)$, then this problem reduces via Floquet theory to the examination of the characteristic exponents associated with (1). In general, this is a hard problem, as is witnessed by the large volume of literature on Mathieu's and and Hill's equations (see, for example [11]).

The condition that f should have the same sign as x for $|x|$ large enough may lead to the result that some continuable solutions with sufficiently large initial conditions

DIFFERENTIAL EQUATIONS

ISBN 0-12-045550-1

have several zeros, or even oscillate. In the linear case, comparison of different solutions via the Sturm separation theorem ensures that different solutions have a comparable zero distribution. For certain types of nonlinearity, on the other hand, the zero distribution is highly dependent on the initial conditions of the solution. This fact can be exploited to demonstrate the existence of periodic solutions as we shall indicate.

Nehari [12] used the calculus of variations to show the existence of infinitely many periodic solutions of period ω, assuming as well as (a), the conditions

(b) $xf(t,x) > 0$ for $x \neq 0$.

(c) $x^{-1-\varepsilon}f(t,x)$ is monotone increasing in x for $x \geq 0$, for some $\varepsilon > 0$.

(d) $f(t,x) = -f(t,-x)$

Note that (c) implies the following:

(e) $\lim_{|x| \to \infty} \frac{f(t,x)}{x} = \infty$, uniformly in t; $\frac{f(t,x)}{x}$ is bounded for $|x| \leq 1$, uniformly in t.

We may regard (e) as prescribing a uniform superlinear behavior for $f(t,x)$ near $|x| = \infty$. Jacobowitz [9] showed that (a), (b) and (e) are sufficient for the existence of infinitely many periodic solutions of (1), and Hartman [8] obtained the same result without requiring the strong sign condition (b). The author considered (1) under an assumption that f was sublinear near $x = 0$, that is, the condition

(f) $\lim_{x \to 0} \frac{f(t,x)}{x} = \infty$, uniformly in t

and showed that (a), (b) and (f) imply infinitely many periodic solutions of (1) [4].

These above results are all based on a fixed point theorem due to Poincaré [13] and Birkhoff [1] which states the following:

POINCARÉ-BIRKHOFF "TWIST" THEOREM

Let A be a bounded topological annular region of $\mathbb{R}^2$ which includes its outer boundary S_1, but not its inner boundary S_0. Let T be a homeomorphism of the closure $\bar{A}$ of A in $\mathbb{R}^2$ such that

(i) $T(S_0) = S_0$

(ii) T preserves Lebesgue measure

(iii) T is a "twist" map of $\bar{A}$.

Then T has a fixed point in A.

Here a "twist" map is, loosely speaking, a map which "twists" S_0 and S_1 in opposite directions. More precisely, let $\tilde{A} = \mathbb{R} \times (0,r_0]$ where $r_0 > 0$. $\tilde{A}$ serves as a universal cover for A, which we may identify with $[0,2\pi) \times (0,r_0]$, and we may lift the restriction of T to A to a periodic homeomorphism $\tilde{T}$ of $\tilde{A}$ on to $\tilde{T}(\tilde{A})$. For $(\theta,r) \in \tilde{A}$, let $\tilde{T}(\theta,r) = (g_1(\theta,r), g_2(\theta,r))$. Then T is said to be a "twist" map if

$$\lim_{r \to 0} (g_1(\theta,r)-\theta) > 0 \quad \text{for all } \theta \text{, and}$$
$$(< 0)$$
$$g_2(\theta,r_0) - \theta < 0 \quad \text{for all } \theta .$$
$$(> 0)$$

Of course $\tilde{T}$ is itself a "twist" homeomorphism of $\tilde{A}$ and it preserves some nonsingular measure induced by the covering map of $\tilde{A}$ to A. The argument to obtain the "twist" theorem is to show first that $\tilde{T}$ has a fixed point in $\tilde{A}$ from which the existence of a fixed point of T immediately follows.

The full assertion of the "twist" theorem is that, in fact, T has at least two fixed points in A. Although, there was for many years some question concerning the validity of Birkhoff's proof of the existence of the second fixed point, the matter was recently affirmatively settled by Brown and Neuman [2].

Returning to the discussion of (1), we sketch the method by which the "twist" theorem may be used to show the existence of periodic solutions.

Write (1) as the system

$$\begin{cases} x' = y \\ y' = -f(t,x) \end{cases} \tag{2}$$

and let A be a disc of radius r_o, punctured at the origin (its centre).

Let $\phi(t;a,b)$ be the solution at the time t of (2) which satisfies $x(0) = a$, $y(0) = b$, and define P on $\bar{A}$ by $P(a,b) = \phi(\omega;a,b)$. P is a homeomorphism of $\bar{A}$ with $P(0,0) = (0,0)$, and on account of the absence of a term involving $x'(t)$ in (1), it is easy to show that P preserves Lebesgue measure. Convert (2) to a system in the polar coordinates (θ,r) in the usual manner and identify the universal cover $\tilde{A}$ of A as a subset of (θ,r)-space. Then P lifts to a homeomorphism $\tilde{P}$ of $\tilde{A}$, and $\tilde{P}$ yields information about the number of zeros of solutions of (1) in the t-interval $[0,\omega)$. Let $\tilde{R}$ be the translation of (θ,r)-space given by $\tilde{R}(\theta,r) = (\theta + 2\pi,r)$.

Now define $\tilde{T}$ to be $\tilde{R}^N \circ \tilde{P}$ for some natural number N. $\tilde{T}$ is a homeomorphism of $\tilde{A}$ with the same measure-preserving properties as $\tilde{P}$ (it preserves the measure $r dr d\theta$). If N and the radius r_0 of A are chosen to be sufficiently large, either of the superlinear or sublinear conditions of f (conditions (e) and (f)) may be used to show that $\tilde{T}$ is a "twist" map. This reflects the fact that these nonlinear conditions may be used to obtain solutions of (1) with a large number of zeros on $[0,\omega)$, provided the initial conditions are chosen suitably. Thus, $\tilde{T}$ has a fixed point in A, and this corresponds to a periodic solution of (1) which has precisely 2N zeros on $[0,\omega)$.

In order to successfully complete this argument, there are two technical problems to overcome. In the case that f is superlinear, there is the problem of extendability of solutions of (1) to $[0,\omega]$ which is necessary to define the maps P and $\tilde{P}$. In [9] this was avoided by using a "first return" map instead of the map P, at the expense of requiring the strong sign condition (b). Hartman overcame this problem of extendability of solutions by employing a suitable "truncation" of the function f. In the case that f is sublinear, there is the problem of uniqueness of solutions of initial value problems; some conditions guaranteeing this are discussed in [4].

Next, we present a refinement of the results that we have referred to above on the existence of ω-periodic solutions. We relax the conditions (e) and (f) concerning the specific nonlinear behavior of $f(t,x)$, requiring only that the appropriate condition should hold for a non-null set of values of t.

Theorem 1. Let $f(t,x)$ satisfy condition (a), that is $f(t,x)$ is continuous, periodic in t with period $\omega > 0$, has the same sign as x for $|x|$ sufficiently large (independently of t), $f(t,0) \equiv 0$ and $f(t,x)$ is sufficiently smooth for the uniqueness of solution of initial value problems of (1).

Suppose, in addition, that there is a measurable set S of positive measure such that either

(e') $$\lim_{|x| \to \infty} \frac{f(t,x)}{x} = \infty \quad \text{for} \quad t \in S;$$

$$\frac{f(t,x)}{x} \text{ is bounded for } 0 < |x| \leq 1,$$

uniformly in t,

or

(f') $$\lim_{x \to 0} \frac{f(t,x)}{x} = \infty \quad \text{for} \quad t \in S;$$

$$\frac{f(t,x)}{x} \text{ is bounded for } |x| \geq 1,$$

uniformly in t.

Then (1) has infinitely many periodic solutions of period ω. For each sufficiently large natural number N, there is such a periodic solution with precisely $2N$ zeros on $[0,\omega)$.

Proof. We shall sketch the proof for the superlinear case (condition (e')) noting that the sublinear case requires a straightforward analogue of the same ideas. The complete details will appear elsewhere [5].

Again the idea is to work with the map $\tilde{T} = \tilde{R}^N \circ \tilde{P}$ which we described previously in our discussion of the way in which the "twist" theorem is used to obtain periodic solutions of (1). What is required here is to show how the condition (e') guarantees that $\tilde{T}$ is a "twist" map for large enough N. Henceforth, we shall assume that conditions (a) and (e') hold, and for simplicity, we shall assume that all solutions of (1) are extendable to the real line. When this extendibility condition is in doubt, the argument may be modified by employing a suitable "truncation" of $f(t,x)$ as utilized in [8]. The key lemmas are

Lemma 1. Let $K > 0$. Then there exists $M = M(K) > 0$ and $t^* = t^*(K) \in [0,\omega)$ such that if $a(t)$ is defined to be $\inf_{|x| \geq M} \frac{f(t,x)}{x}$, then $a(t)$ is a nonnegative measurable (locally integrable) function such that

$$\int_s^t a(\sigma)d\sigma > \frac{Kct}{2\omega} - Ks \quad \text{whenever} \quad t^* \leq s \leq t ,$$

where c is the measure of $S \cap [0,\omega)$.

Lemma 2. Let N be a positive integer. Then there exists $t_0 = t_0(N) \in [0,\omega)$ and $R = R(N) > 0$ such that for any solution $x(t)$ of (1) satisfying $x^2(t_0) + x'^2(t_0) \geq R^2$, x has at least N zeros on $[t_0, t_0 + \omega)$.

Lemma 2 is somewhat similar to Lemma 3 of [9] and Lemma 2.1 of [8]; the important difference being that instead of using the Sturm comparison theorem as part of the proof, we use two integral comparison theorems of Levin [10], [14, Theorems 1.35 and 1.36]. This is where Lemma 1 is employed.

Lemma 3. There is a constant N_0 such that if $x(t)$ is any solution of (1) satisfying $x^2(t_0) + x'^2(t_0) \leq 1$ for any $t_0 \in [0,\omega)$, then x has at most N_0 zeros on $[0,\omega)$.

Lemmas 1-3 enable the appropriate "twist" map to be exhibited, and the rest of the proof of the theorem is straightforward.

In [8], it was pointed out that the equation

$$x'' + f(t,x,x') = 0$$

does not in general have nontrivial periodic solutions. However, in certain cases equations with damping terms do exhibit periodic solutions. Thus, we have the following

Theorem 2. Let $f(t,x)$ be as in Theorem 1 and let $q(t)$ be a continuous ω-periodic function with $\int_0^\omega q(s)ds = 0$.

Then the equation

$$x''(t) + q(t)x'(t) + f(t,x(t)) = 0 \tag{3}$$

has infinitely many periodic solutions of period ω.

Proof. We may write (3) as

$$(r(t)x'(t))' + r(t)f(t,x(t)) = 0 \tag{4}$$

where $r(t) = \exp\left(\int_0^t q(s)ds\right)$.

The conditions on q imply that $\int_0^\infty \frac{du}{r(u)} = \infty$ and so we may transform (4) by putting $T = \int_0^t \frac{du}{r(u)}$, $y(T) = x(t)$ to obtain

$$\frac{d^2y}{dT^2} + r^2(t(T))f(t(T),y(T)) = 0 \tag{5}$$

Denoting $r^2(t(T))f(t(T),y)$ by $g(T,y)$, it is easily verified that $g(T,y)$ is periodic in T with period

$\Omega = \int_0^{\omega} \frac{du}{r(u)}$, and one sees immediately that $g(T,y)$ satisfies condition (a) (with Ω in place of ω) and that $\frac{g(T,y)}{y}$ is bounded for $|y| \le 1$, independently of T, or is bounded for $|y| \ge 1$, independently of T, according to whether f satisfies condition (e') or (f'). Finally, $T(t)$ is a diffeomorphism and so $T(S)$ is a measurable set of positive measure such that $\lim_{|y|\to\infty} \frac{g(T,y)}{y} = \infty$ for $T \in T(S)$ (if f satisfies condition (e')) or $\lim_{y\to 0} \frac{g(T,y)}{y} = \infty$ for $T \in T(S)$ (if f satisfies condition (f')).

Thus g satisfies the hypotheses of Theorem 1, implying that (5) has infinitely many periodic solutions of period Ω, and so (4) has infinitely many periodic solutions of period ω.

Remarks. 1. A result similar to that of Theorem 2 is available in some situations where the sign condition on $f(t,x)$ for large $|x|$ is relaxed, based on the results of [3]. For example, suppose that $p(t)$, $q(t)$ are ω-periodic continuous functions such that p is of locally bounded variation, has isolated zeros and changes sign at least once on $[0,\omega)$, and $\int_0^{\omega} q(t)dt = 0$. Suppose that $g(x)$ is locally lipschitz with $xg(x) \ge 0$,

$$\lim_{|x|\to\infty} \frac{g(x)}{x} = \infty \quad \text{and} \quad \int_0^{\pm\infty} [1 + G(s)]^{-\frac{1}{2}} ds < \infty ,$$

where $G(X) = \int_0^x g(u)du$. Then the equation

$$x''(t) + q(t)x'(t) + p(t)g(x(t)) = 0$$

has infinitely many ω-periodic solutions.

2. One might expect that in Theorem 1, we cannot allow the set S of values of t for which $\lim_{|x|\to\infty} \frac{f(t,x)}{x} = \infty$ (or $\lim_{x\to 0} \frac{f(t,x)}{x} = \infty$) to be too small and still retain the validity of the theorem. We give the following example that shows that if $\lim_{|x|\to\infty} \frac{f(t,x)}{x} = \infty$ holds only for a single point $t \in [0,\omega)$, then there may not be any nontrivial ω-periodic solutions: Consider the solution $y(t)$ of the equation $y'' + y^3 = 0$ for which $y(0) = 0$, $y'(0) = B > 0$. $y(t)$ attains its maximum value of $2^{\frac{1}{4}}B^{\frac{1}{2}}$, at a time $T(B) = kB^{-\frac{1}{2}}$, where $k = \frac{1}{4} \cdot 2^{\frac{1}{4}} \int_0^1 \frac{dv}{(1 - v^4)^{\frac{1}{2}}}$.

Define the function $g(t)$ as follows:

$$g(t) = \begin{cases} \dfrac{2^{\frac{1}{4}}k}{t}, & 0 < t \le \dfrac{1}{2} \\[2ex] \dfrac{2^{\frac{1}{4}}k}{1-t}, & \dfrac{1}{2} < t < 1 \end{cases}$$

$g(t)$ will be undefined for integer values of t and defined for all other values of t so that it is periodic of period 1. Thus, if $y(t)$ solves $y'' + y^3 = 0$, $y(0) = 0$, $y'(0) = B > 0$, then $y(T(B)) = g(T(B))$, $y'(T(B)) = 0$, provided $T(B) < \frac{1}{2}$.

Let $\mathbb{Z}$ denote the set of integers and define $f(t,x)$ as follows:

$$f(t,x) = \begin{cases} x^3 , \ 0 \le x \le \frac{1}{2}\, g(t) , \ t \notin \mathbb{Z} \\ \frac{2(g(t) - x)}{g(t)}\, x^3 , \ \frac{1}{2}\, g(t) < x \le g(t), \ t \notin \mathbb{Z} \\ 0, \ x > g(t), \ t \notin \mathbb{Z} \\ x^3 , \ t \in \mathbb{Z} \end{cases}$$

For $x < 0$, we define $f(t,x)$ to be $-f(t,-x)$. f is a continuous function of period 1 in t and satisfies condition (a). We note that

$$\lim_{|x| \to \infty} \frac{f(t,x)}{x} = \begin{cases} \infty , & t \in \mathbb{Z} \\ 0 , & t \notin \mathbb{Z} . \end{cases}$$

It is not hard to verify that there is no nontrivial solution of

$$x''(t) + f(t,x(t)) = 0 \tag{6}$$

which has more than one zero on $[0,1)$. Since $xf(t,x) \ge 0$ for all x, any periodic solution of (6) must have at least two zeros on $[0,1)$. Thus, (6) has no nontrivial solutions of period 1.

It is, however, true that (6) does have periodic solutions of a sufficiently large integer period. We do not know whether this example may be modified so that there are no periodic solutions of any period.

3. There are a large number of papers devoted to the study of the equation (1) modified by a periodic forcing term,

that is

$$x''(t) + f(t,x(t)) = p(t)$$

(see for example [6,7]).

Probably Theorem 1 can be extended to such equation, at least in the superlinear case.

REFERENCES

[1] Birkhoff, G. D., *Proof of Poincaré's geometric theorem,* Trans. Amer. Math. Soc. 14(1913), pp. 14-22.

[2] Brown, M., and W. D. Neumann, "*Proof of the Poincaré-Birkhoff fixed point theorem,* Michigan Math. J., 24(1977), pp. 21-31.

[3] Butler, G. J., "*Rapid oscillation, nonextendability and the existence of periodic solutions to second order non-linear ordinary differential equations,* J. Differential Equations, 22(1976), pp. 467-477.

[4] Butler, G. J., *Periodic solutions of sublinear second order differential equations,* J. Math. Anal. Appl., 62 (1978), pp. 676-690.

[5] Butler, G. J., *On boundary value problems for nonlinear second order differential equations and a perturbed version of the "twist" theorem,* (in preparation).

[6] Chang, S. H., *Existence of periodic solutions to second order nonlinear equations,* J. Math. Anal. Appl, 52(1975), pp. 255-259.

[7] Fucik, S. and V. Lovicar, *Periodic solutions of the equation* $x''(t) + g(x(t)) = p(t)$, Casopis pest. mat. 100 (1975), pp. 160-175.

[8] Hartman, P., *On boundary value problems for superlinear second order differential equations*, J. Differential Equations, 26(1977), pp. 37-53.

[9] Jacobowitz, H., *Periodic solutions of* $x'' + f(x,t) = 0$ *via the Poincaré-Birkhoff Theorem*, J. Differential Equations, 20(1976), pp. 37-52.

[10] Ju Levin, A., *A comparison principle for second-order differential equations*, Soviet Math. Dokl. 1(1960), pp. 1313-1316.

[11] Magnus, W., and S. Winkler, "Hill's Equation," John Wiley and Sons, New York, 1966.

[12] Nehari, Z., *Characteristic values associated with a class of nonlinear second order differential equations*, Acta Math., 105(1961), pp. 141-175.

[13] Poincaré, H., *Sur un théorème de géométrie*, Rend. Circ. Mat. Palermo 33(1912), pp. 375-407.

[14] Swanson, C. A., "Comparison and Oscillation Theory of Linear Differential Equations," Academic Press, New York, 1968.

PERIODIC SOLUTIONS OF THE FORCED PENDULUM EQUATION

Alfonso Castro

C.I.E.A. del I.P.N.
Apartado Postal 14740
México 14, D.F. MEXICO

INTRODUCTION

Here we study the existence of 2π-periodic weak solutions for the equation

$$x'' + g(x(t)) = p(t), \qquad (1.1)$$

where $p : \mathbb{R} \to \mathbb{R}$ is a 2π-periodic measurable function with $\int_0^{2\pi} p^2 < \infty$ and $g : \mathbb{R} \to \mathbb{R}$ is a continuous T-periodic function such that

$$(g(u) - g(v))/(u - v) < 1 \text{ for all } u, v, \in \mathbb{R}, \; u \neq v. \qquad (1.2)$$

For example, the classical pendulum equation where $g(u) = \sin(u)$ satisfies (1.2).

Our main goal is to prove:

<u>Theorem A</u>: <u>Suppose</u> (1.2) <u>holds</u>. <u>Let</u> $\bar{g} = (1/T)\int_0^T g(s)ds$. <u>If</u> $p_0 : \mathbb{R} \to \mathbb{R}$ <u>is a</u> 2π-<u>periodic measurable function such that</u> $\int_0^{2\pi} p_0 = 0$ <u>and</u> $\int_0^{2\pi} p_0^2 < \infty$ <u>then there exist two real numbers</u> $d(p_0)$ <u>and</u> $D(p_0)$, <u>with</u>

$$\min_{t \in \mathbb{R}} g(t) \leq d(p_0) \leq \bar{g} \leq D(p_0) \leq \max_{t \in \mathbb{R}} g(t) \;, \qquad (1.3)$$

ISBN 0-12-045550-1

such that for $p_1 \in \mathbb{R}$ *the equation*

$$x'' + g(x(t)) = p_0(t) + p_1 \tag{1.4}$$

has a 2π-*periodic weak solution (therefore infinitely many) iff* $d(p_0) \le p_1 \le D(p_0)$. *In addition,*

i) if $C = \{x \mid g(x) = \max_{s \in \mathbb{R}} g(s), \text{ or, } g(x) = \min_{s \in \mathbb{R}} g(x)\}$ *is discrete then*

$$d(p_0) = \min_{t \in \mathbb{R}} g(t) \quad \text{or} \quad D(p_0) = \max_{t \in \mathbb{R}} g(t), \tag{1.5}$$

iff, $p_0 = 0$, *and*

ii) *if* $\{p_0^n\}$ *is a sequence converging weakly in* $L_2[0,2\pi]$ *to* p_0 *then* $d(p_0^n) \to d(p_0)$ *and* $D(p_0^n) \to D(p_0)$.

For other results where the range of a nonlinear operator is completely characterized see [3] and [10]. Recent developments concerning the range of the sum of two operators do not apply here. In fact, in particular, Theorem A shows that this is not a case where $R(A + B) \simeq R(A) + R(B)$ or $R(A + B) \simeq R(A) + \text{conv}(R(B))$ (see [4], [5]) .

We want to point out a few remarks relating Theorem A with Theorem 4.1 of [2] . First, Theorem A cannot be derived from results of [2] because (1.2) is a much weaker condition than the assumption of [2]

$$g' \le \text{const.} < 1 . \tag{1.6}$$

In fact, we can allow g' to take the value 1. Second, Theorem 4.1 of [2] does not guarantee the existence of solutions of (1.4) when p_1 is either $d(p_0)$ or $D(p_0)$. Third, since the methods used here are variational, additional information on the stability of the solution can be obtained.

The inequality (1.3) answers in the affirmative a question posed in [11]. Since $g \equiv$ constant satisfies (1.2) and, for such a g, $d(p_0) \equiv D(p_0) \equiv \bar{g}$, the inequality (1.3) is sharp. To the best of our knowledge it is an open question whether $g \neq$ constant implies $d(p_0) < D(p_0)$, for every p_0. From the proof of Theorem A, it follows that if $d(p_0) = D(p_0)$, then for each constant function x there exists a function $\varphi(x,p_0)$, with $\int \varphi(x,p_0) = 0$, such $x + \varphi(x,p_0)$ is a solution of $u'' + g(u) = p_0 + d(p_0)$.

2. NOTATIONS AND PRELIMINARY LEMMAS

All integrals will be over the invertal $[0,2\pi]$ unless otherwise indicated.

We let H be the Sobolev space of 2π-periodic functions $u : \mathbb{R} \to \mathbb{R}$ with $u \in L_2[0,2\pi]$ and generalized first order derivative $u' \in L_2[0,2\pi]$. The inner product in H is given by the bilinear form

$$[u,v] = (1/4\pi^2)(\int u(t)dt)(\int v(t)dt) + \int u_0'(t)v_0'(t)dt,$$

where $u_0(t) = u(t) - (1/2\pi)\int u(t)dt$ and $v_0(t) = v(t) - (1/2\pi)\int v(t)dt$. We let X denote the subspace of H generated by the constant functions and we denote by Y the subspace H of all the functions with mean value zero. It is easy to verify that

$$\int y^2 \leq \int (y')^2 \quad \text{for all } y \in Y . \tag{2.1}$$

For each 2π-periodic measurable function $p : R \to R$ such that $\int p^2 < \infty$ we let $J_p : H \to \mathbb{R}$ be the functional defined by

$$J_p(u) = \int((u'(t))^2/2 - G(u(t)) + p(t)u(t))dt,$$

where $G(u) = \int_0^u g(s)ds$. Since g is continuous, J_p is of class C^1 and for each $u, v \in H$

$$[\nabla J_p(u), v] \equiv \lim_{t \to 0} \frac{J_p(u + tv) - J_p(u)}{t} \tag{2.2}$$

$$= (u'(t)v'(t) - g(u(t))v(t) + p(t)v(t))dt$$

It is clear that $u_0 \in H$ is a critical point of J_p iff u_0 is a weak solution of (1.1). We denote $p_0 = p - (1/2\pi)p$ and $p_1 = p - p_0$.

<u>Lemma 1.</u>: <u>Let</u> p, p_0, p_1, X, Y <u>and</u> J_p <u>be as above</u>. <u>For each</u> $x \in X$ <u>there exists a unique</u> $\phi(x,p) \in Y$ <u>such that</u>

$$[\nabla J_p (x + \phi(x,p)), y] = 0 \text{ for all } y \in Y. \tag{2.3}$$

<u>Moreover, the function</u> $\phi(\ ,p) : X \to Y, x \to \phi(x,p)$ <u>is continuous</u>.

<u>Proof</u>: Suppose $y_1, y_2 \in Y$ are such that

$$[\nabla J_p (x + y_1), y] = 0 = [\nabla J_p(x + y_2), y] \tag{2.4}$$

for all $y \in Y$. Thus, $[\nabla J_p(x + y_1) - \nabla J_p(x + y_2), y_1 - y_2] = 0$, and we have

$$\int(((y_1 - y_2)'(t))^2 - (g(x + y_1(t)) - g(x + y_2(t))) ((y_1(t) - y_2(t)))dt = 0. \tag{2.5}$$

The relations (1.2), (2.1) and (2.5) imply that $y_1 = y_2$. Hence, the uniqueness of $\phi(x,p)$ is proved.

Since the inclusion of H in $L_2[0,2\pi]$ is compact (see [1,th.6.2.]) and $J_p(x + y) = (\|y\|^2)/2 - \int_0^{2\pi} (G(x + y(t)) - p(t)(x + y(t)))dt$ it follows that J_p restricted to the linear manifold $\{x + y; y \in Y\}$ attains a point of minimum. Thus, this point of minimum satisfies the relation of (2.3). Consequently, we have proved the existence of $\phi(x,p)$.

The boundedness of g and (2.5) implies that $\phi(x,p)$ is bounded in H. This and the uniqueness of $\phi(x,p)$ imply that $\phi(\ ,p)$ is a continuous function and the proof of lemma 1 is complete.

Suppose that c is any constant function. From (2.3) we have

$$[\nabla J_{p+c}(x + \phi(x,p)), y] \tag{2.6}$$
$$= \int (\phi(x,p))'y' - g(x + \phi(x,p))y + (p + c)y$$
$$= \int (\phi(x,p))'y' - g(x + \phi(x,p))y + py = 0$$

for all $y \in Y$. From (2.6) and the definition of $\phi(\ ,p + c)$ we infer that $\phi(\ ,p) \equiv \phi(\ ,p + c)$. Thus, we have proved:

Lemma 2. *If* p, p_0, p_1 *and* $\phi(\ ,p)$ *are as in lemma* 1 *then* $\phi(\ ,p)$ *depends only on* p_0.

We denote by $\tilde{J}_p : X \to \mathbb{R}$ the function defined by $\tilde{J}_p(x) = J(x + \phi(x,p))$. Since J is of class C^1 and $\phi(\ ,p)$ is continuous, following the ideas of lemma 2.1 of [6] it is easy to prove that $\tilde{J}$ is a class C^1 and that

$$[\nabla\tilde{J}_p(x), x_1] = [\nabla J_p(x + \phi(x,p)), x_1] \tag{2.7}$$

Lemma 3.: *The element* $x + y \in X \oplus Y$ *is a critical point of* J_p *iff* $y = \phi(x,p)$ *and* x *is a critical point of* $\tilde{J}_p$.

Proof: =>) By lemma 1 we see that if $x + y \in X \oplus Y$ is a critical point of J_p then $y = \phi(x,p)$. This and (2.7) imply that x has to be a critical point of $\tilde{J}_p$.

<=) Suppose $y = \phi(x,p)$ and $\nabla\tilde{J}_p(x) = 0$. Let $u \in H$. Since $H = X \oplus Y$ we can write $u = x_1 + y_1$, with $x_1 \in X$ and $y_1 \in Y$.

Thus,

$$[\nabla J_p(x + y),\ u] = [\nabla J_p(x + \phi(x,p)), y_1] +$$
$$[\nabla J_p(x + \phi(x,p)),\ x_1] = 0 + [\nabla \tilde{J}_p\ (x), x_1] = 0.$$

This proves lemma 3.

Lemma 4.: The function $\phi(\ ,p)$ is T-periodic.

Proof: Let $x \in X$,

$$\begin{aligned}[\nabla J_p(x + T + \phi(x,p)), y] &= \int (\phi(x,p))'\ y' - g(x + T + \phi(x,p))y + py \\ &= \int (\phi(x))'\ y' - g(x + \phi(x,p))\ y + py = 0\end{aligned}$$

for all $y \in Y$. Thus, by lemma 1, $\phi(x + T, p) = \phi(x, p)$ and lemma 4 is proved.

3. PROOF OF THEOREM A

For each $p_0 : \mathbb{R} \to \mathbb{R}$ 2π-periodic measurable function such that $\int p_0 = 0$ and $\int p_0^2 < \infty$ let

$$d(p_0) = (1/2\pi)\ \min \int (g(x + \phi(x,p_0)(t)))dt \tag{3.1}$$
$$D(p_0) = (1/2\pi)\ \max \int (g(x + \phi(x,p_0)(t)))dt\ .$$

According to (2.7) and lemma 2, the equation (1.4) has a 2π-periodic weak solution iff there exists $x \in X$ such that for every $x_1 \in X$ $[\nabla \tilde{J}_p\ (x), x_1] = 0$. Thus, from (2.2) and (2.7), we see that (1.4) has a 2π-periodic weak solution iff there exists $x \in X$ such that

$$\begin{aligned} 0 &= \int \phi(x,p_0)'\ x_1' - g(x + \phi(x,p_0))x_1 + p_0 x_1 + p_1 x_1 \\ &= x_1 \int (-g(x + \phi(x,p_0)(t)) + p_1)dt \end{aligned} \tag{3.2}$$

for every constant function x_1. From (3.1) and (3.2) we see immediately that (1.4) has a 2π-periodic weak solution iff $d(p_0) \le p_1 \le D(p_0)$.

In order to prove that $d(p_0) \le \overline{g} \le D(p_0)$ we first observe that the function $G(s) - \overline{g}s$ is T-periodic. Hence, setting $p_1 = \overline{g}$, we see that

$$\tilde{J}(x) = \int (((\varphi(x,p_0)')^2/2 - G(x + \varphi(x,p_0)) - \overline{g}(x + \varphi(x,p_0)) + p_0\, \varphi(x,p_0))$$

is T-periodic and differentiable. Therefore $\tilde{J}$ has infinitely many critical points. Thus, by lemma 3, the equation $u'' + g(u) = p_0 + \overline{g}$ has infinitely many 2π-periodic solutions and the inequality (1.3) is proven.

Now we prove part i). Since $\phi(x, 0) = 0$ for every $x \in X$, from (3.1) we have

$$d(0) = (1/2\pi) \min_{x \in X} \int g(x)\, dt = \min_{t \in R} g(t) .$$

Similary, $D(0) = \max\{g(t); t \in R\}$. Suppose $p_0 \ne 0$, we claim that $\phi(x,p_0) \ne 0$ for every $x \in X$. In fact, if for some $x \in X$ $\phi(x,p_0) = 0$ then, by (2.3), we have

$$0 = \int -g(x)y + p_0 = \int p_0 y \quad \text{for all } y \in Y .$$

But this relation contradicts that $p_0 \ne 0$. Since C is discrete, for each $x \in X$ there exist $t \in [0,2\pi]$ such that $\min\{g(s); s \in R\} < g(x + \phi(x,p_0)(t)) < \max\{g(s); s \in R\}$. Therefore, for each $x \in X$

$$\min\{g(s); s \in R\} < (1/2\pi) \int g(x + \phi(x,p_0)(t))dt = \max\{g(s); s \in R\} . \tag{3.3}$$

In the last set of inequalities we have used that every element in H is a continuous function (see [1,th.5.4.]).

Since $\phi(\ ,p_0)$ is continuous we have that the function $f : X \to \mathbb{R}$, $x \to \int g(x + \phi(x,p_0))dt$ is continuous. This and (3.3) conclude part i) of the theorem.

Suppose $\{p_0^n\}$ is a sequence of 2π-measurable functions converging weakly in $L_2[0,2\pi]$ to p_0 . We denote $J_n = J_{p_0^n}$, $J = J_{p_0}$, $\phi_n = \phi(\ ,p_0^n)$ and $\phi = \phi(\ ,p_0)$. Since the function g is bounded and $\{p_0^n\}$ is bounded in $L_2[0,2\pi]$, from

$$0 = \int((\phi_n(x))')^2 - (g(x + \phi_n(x)) + p_0^n)\phi_n(x) \tag{3.4}$$

we obtain that $\{\phi_n(x);\ n = 1, 2, \ldots, x \in X\}$ is bounded. Let $x_0 \in X$ be fixed. Let $\psi \in Y$ be such that some subsequence $\{\phi_{n_j}(x_0)\}$ converges weakly to ψ . Therefore, since for every $y \in Y$

$$0 = \int(\phi_{n_j}(x_0))'y' - g(x_0 + \phi_{n_j}(x_0))y + p_0^{n_j}y$$

and $\phi_{n_j}(x_0)$ converges in $L_2[0,2\pi]$ to ψ (see [1,th.6.2]), we have

$$0 = \int \psi'y' - g(x_0 + \psi)y + p_0 y \quad \text{for all } y \in Y . \tag{3.5}$$

From (3.5) and lemma 1 we see that $\psi = \phi(x_0)$. Also, we obtain that $\phi_n(x_0)$ converge weakly in H to $\phi(x_0)$.

We prove now that $D(p_0^n) \to D(p_0)$. Let $u_n \in H$ be a weak solution of

$$u'' + g(u) = p_0^n + D(p_0^n) . \tag{3.6}$$

Since g is periodic we can assume that for every $n \in N$ the sequence of projections of the u_n's on X is bounded. Thus, by lemma 3 and (3.4) we can assume that $\{u_n\}$ is bounded in H. Let $u_0 \in H$ be such that some subsequence $\{u_{n_j}\}$

converges weakly to $u_0 \in H$. This and (3.6) imply

$$\int u_0'v' - g(u_0)v + p_0 v + (\limsup D(p_0^{n_j}))v = 0$$

for all $v \in H$. This proves that $D(p_0) \geq \limsup D(p_0^{n_j})$ and, therefore,

$$D(p_0) \geq \limsup (D(p_0^n)) . \tag{3.7}$$

Let $x_1 \in X$ be such that $D(p_0 = (1/2\pi) \int g(x_1 + \phi(x_1)(t))dt$. From (3.5) we know that $\phi_n(x_1)$ converges in $L_2[0, 2\pi]$ to $\phi(x_1)$. Thus,

$$D(p_0) = (1/2\pi) \lim_n \int g(x_1 + \phi_n(x_1)(t))dt .$$

Since $\int g(x_1 + \phi_n(x_1)(t))dt \leq 2\pi D(p_0^n)$, we have $D(p_0) \leq \liminf D(p_0^n)$. This combined with (3.7) proves that $d(p_0^n) \to d(p_0)$. In a similar manner it can be proved that $d(p_0^n) \to d(p_0)$, and the proof of the theorem is complete.

4. AN ELLIPTIC PROBLEM WITH NEUMANN BOUNDARY CONDITIONS

The above developments can be applied to the Neumann problem

$$\Delta u + g(u) = p(x) \qquad x \in \Omega \tag{4.1}$$

$$(\partial u/\partial \eta)(x) = 0 \qquad x \in \partial\Omega$$

if we assume that

a) Ω is a bounded region in $\mathbb{R}^n$ with smooth boundary,

b) $p \in L_2(\Omega)$,

and c) g is periodic function of class C^1 such that $g'(t) \leq \lambda_1$ and $\{t \in \mathbb{R} ; g'(t) \in \{0,\lambda_1\}\}$ is discrete.

In the above statements $(\partial u/\partial \eta)$ denotes derivative of u in the outer normal direction on $\partial\Omega$ and λ_1 denotes the first positive eigenvalue of

$$\Delta u + \lambda u = 0 \qquad \Omega$$

$$(\partial u / \partial \eta) = 0 \qquad \partial \Omega$$

The reader may verify the following

<u>Theorem B.</u>: <u>Each function</u> $p_0 \in L_2(\Omega)$ <u>such that</u> $\int_\Omega p_0 = 0$ <u>determines two numbers</u> $d(p_0)$ <u>and</u> $D(p_0)$ <u>which satisfy</u>:

i) <u>If</u> $p_1 \in \mathbb{R}$ <u>then the Neumann problem</u> $\{\Delta u + g(u) = p_0 + p_1$ in Ω, $\partial u / \partial \eta \equiv 0$ on $\partial \Omega\}$ <u>has a weak solution iff</u> $d(p_0) \le p_1 \le D(p_0)$,

ii) $\min_{t \in R} g(t) \le d(p_0) \le \overline{g} \le D(p_0) \le \min_{t \in R} g(t)$,

iii) $d(p_0) = \min_{i \in \mathbb{R}} g(t)$ <u>or</u> $D(p_0) = \min_{t \in \mathbb{R}} g(t)$, <u>iff</u>, $p_0 = 0$ <u>and</u>

iv) <u>if</u> p_0^n <u>converges weakly in</u> $L_2(\Omega)$ <u>to</u> p_0 <u>then</u> $d(p_0^n) \to d(p_0)$ <u>and</u> $D(p_0^n) \to D(p_0)$.

ACKNOWLEDGMENTS

The author is greatly indebted to professors A. C. Lazer and D. de Figueiredo for their helpful comments and remarks. Part of this research was done while the author was visiting the University of Brasilia.

REFERENCES

[1] Adams, R., "Sobolev Spaces," Academic Press, New York (1975).

[2] Ambrosetti, A. and G. Mancini, *Existence and multiplicity results for nonlinear elliptic problems with linear part at resonance. The case of simple eigenvalue.* J. Differential Equations 28(1978), pp. 220-245.

[3] Ambrosetti, A. and G. Prodi, *On the inversion of some differentiable mappings with singularities between Banach Spaces*, Ann. Math. Pura Appl., 93(1972), pp. 231-246.

[4] Brezis, H. and A. Haraux, *Image d'une somme d'operateurs monotones et applications*, Israel J. Math., 23(1976), pp. 165-186.

[5] Brezis, H. and L. Niremberg, *Characterization of the ranges of some nonlinear operator and applications to boundary value problems*, Ann. Scuola Norm. Sup. Pisa, 2(1978), pp. 225-326.

[6] Castro, A., *Hammerstein integral equations with indefinite kernel*, Internat. J. Math. and Math. Sci., 1 (1978) pp. 187-201.

[7] de Figueiredo, D., *The Dirichlet problem for nonlinear elliptic equations: A Hilbert space approach. Partial differential equations and related topics*, Springer Verlag Lecture Notes 446(1975), pp. 144-165.

[8] Landesman, E. and A. C. Lazer, *Nonlinear perturbations of linear elliptic boundary value problems at resonance*, J. Math. Mech., 19(1970), pp. 609-623.

[9] Lazer, A. C., E. Landesman and D. Meyers, *On saddle point problems in the calculus of variations, the Ritz algorithm, and monotome convergence*, J. Math. Annal. Appl. 52(1975), pp. 594-614.

[10] Podolak, E., *On the range of operator equations with an asymptotically nonlinear term*, Indiana Univ. Math. J. 25(1976), pp. 1127-1137.

[11] Konecny, M., *Remarks on periodic solvability of nonlinear ordinary differential equations*, Commentations Mathematicae Universitatis Carolinae 18,3(1977), pp. 547-562.

ON THE STRUCTURAL IDENTIFICATION (INVERSE) PROBLEM FOR ILLNESS-DEATH PROCESSES

Jerome Eisenfeld

University of Texas at Arlington

INTRODUCTION

The inverse problem may be described as follows [1]. Consider the system of differential equations $\dot{\underline{x}}(t,\theta) = f[\underline{x}(t,\theta),\underline{u}(t),t,\underline{\theta}]$, $\underline{x}(t_0) = \underline{x}_0$, $\underline{y}(t,\underline{\theta}) = h[\underline{x}(t,\underline{\theta}),\underline{u}(t),t,\underline{\theta}]$ where $\underline{x}(t,\underline{\theta}) \in R^n$, $\underline{y}(t,\underline{\theta}) \in R^p$, $\underline{\theta} \in \Omega \subset R^r$, $u(t) \in U \subset D[R_+ \to R^q]$, $f: R^n \times U \times R_+ \times \Omega \to R^n$, $h: R^n \times U \times \Omega \to R^p$. The inverse problem consists of determining the unknown parameter $\underline{\theta}$ by extracting information from observations of $\underline{u}$ and $\underline{y}$. A natural question that arises in the context of inverse problems is whether or not the unknown parameters can be uniquely identified. This is the so-called structural identification problem. The paper discusses necessary and sufficient conditions for structural identification with a view towards application to illness-death processes.

2. Illness-Death Processes. An illness-death process [2-5] may be described in terms of a continuous Markov model with n illness (transient) states each having one (and only one) transfer into a death state. These transfers represent exists due to death or loss to follow-up. Let $x_i(t)$ denote the mean population of patients in illness state i at time

ISBN 0-12-045550-1

t and let $x(t) = (x_1(t), x_2(t), \ldots, x_n(t))^*$ be the vector of mean populations ("*" is the transpose operation). In a simple illness-death process $\underline{x}(t)$ satisfies the differential equation

$$\dot{\underline{x}}(t) = A\underline{x}(t), \quad \underline{x}(0) = \underline{m}, \tag{1}$$

where the matrix element a_{ij} denotes the intensity $(j \to i)$. The death intensities are given by $a_{0j} = -\sum_i a_{ij} > 0$ ($a_{ij} \geq 0$ if $i \neq j$, $a_{ii} \leq 0$).

It is important to note that in a typical study (we have in mind the study in [6]) one does not observe each $x_i(t)$ directly for this would require repeated screening of each patient. However, the total number of survivors,

$$y(t) = x_1(t) + x_2(t) + \ldots + x_n(t), \tag{2}$$

is observed. Other information is also available because each patient's diagnosis is known at $t = 0$. Thus, if we denote by $z_{ji}(t)$ the mean number of patients in state j at time t, which were in state i at $t = 0$, then the functions

$$y_i(t) = z_{1i}(t) + z_{2i}(t) + \ldots + z_{ni}(t), \tag{3}$$

$i = 1,2,\ldots,n$,

are observed. However, it is important to note that the $y_i(t)$ are observed only discretely in time, say at instances $t = h, 2h, \ldots, \nu h$ (we can assume that the integer $\nu \geq n$). Thus, the following mathematical question arises: do the discrete values $y_i(kh)$, $i,k = 1,2,\ldots,n$ (and $\underline{m}$) determine the matrix A uniquely? In other words we have a system of n^2 nonlinear equations involving up to n^2 unknowns and we are asking if the system has a unique solution. In order to deal with this problem some results from control theory are useful.

3. The Bellman-Astrom Theorem and Related Results. Suppose we consider a system of differential equation:

$$\dot{\underline{x}}(t) = A\underline{x}(t) + B\underline{u}(t), \quad \underline{x}(0) = 0, \tag{4}$$

$$\underline{y}(t) = C\underline{x}(t), \tag{5}$$

where $\underline{x}(t)$ is the state variable, $\underline{u}(t)$ is the input, $y(t)$ is the output and A,B,C are respectively the system, input and output matrices. The impulse response function (matrix) for the system is

$$\phi(t) = C\, e^{tA}B.$$

The following result is proved in [7].

Theorem 1. A single-input, completely controllable system is identifiable from the values of its impulse response, obtained on an interval $[0,T]$, if the output matrix is the identity (i.e. all states are observed).

Theorem 1 does not apply to an illness-death process described above because not all states are observed and the observations occur at discrete times.

Applying Theorem 1 to the dual system, (B^*,A^*,C^*), of a system (C,A,B), we obtain immediately the dual result.

Theorem 2. A single-output, completely observable system is identifiable from the values of its impulse response on a finite interval $[0,T]$ if the input matrix is the identity.

Leaving aside discrete observations, Theorem 2 applies to illness-death processes if $\underline{m} \geq 0$. The equations are set into standard form (4)-(5) with $\underline{u}(t) = \delta(t)m$ ($\delta(t)$ is the Dirac delta function), $\underline{y}(t) = (y_1(t), y_2(t), \ldots, y_n(t))$ (defined in (3)) and

$$C = \underline{c} = [1,1,\ldots,1].$$

In other words, the values of $y_i(t)$ in any interval, $[0,T]$, suffice to determine A uniquely proveded each state is nonempty at $t = 0$ and the system is completely observable. The latter condition is obtained if and only if the n vectors $\underline{c}, \underline{c}\,A, \ldots, \underline{c}\,A^{n-1}$ are linearly independent ([8]).

We now return to the problem created by discrete observations. We may assume, without loss of generality, that the sample time $h = 1$, for this requires only a rescaling of the time variable. Now we observe $y_i(t)$, $i,t = 1,2,\ldots,n$. Let $P = e^A$, then it is not difficult to see that the matrix $Z(t)$, with ij element $z_{ij}(t)$, is given by

$$Z(t) = P^t M, \quad t = 1,2,\ldots,n, \tag{6}$$

where $M = \mathrm{diag}(m_1, m_2 \ldots, m_n)$. Moreover,

$$\underline{y}(t) = \underline{c}\,Z(t), \quad t = 1,2,\ldots,n\,. \tag{7}$$

We may then regard (6)-(7) as a discrete system. The analogue of Theorem 2 for the discrete system (with $m_i \neq 0$, $i = 1,2,\ldots,n$) is Lemma 1.

<u>Lemma 1</u>. Given the discrete system (6)-(7) where M is a known nonsingular matrix. Then the discrete values $\underline{y}(t)$, $t = 1,2,\ldots,n$ identifies P uniquely provided the n vectors, $\underline{c}, \underline{c}\,P, \ldots, \underline{c}\,P^{n-1}$ are linearly independent.

<u>Proof</u>: Let $\underline{w}(t) = \underline{y}(t)M^{-1}$, then $\underline{w}(t) = \underline{c}\,P^t$, $t = 1,2,\ldots,n$. Clearly $\underline{w}(t+1) = \underline{w}(t)P$, $t = 0,1,\ldots,n = 1$, where, for convenience of notation, we set $\underline{w}(0) = \underline{c}$. The above n equations may be expressed more compactly in matrix

form:

$$[\underline{w}(1)^*,\underline{w}(2)^*,\ldots,\underline{w}(n)^*] = P^*[\underline{w}(0)^*,\underline{w}(1)^*,\ldots,\underline{w}(n-1)^*].$$

By the linear independence condition, the matrix on the right is nonsingular, thus

$$P^* = [\underline{w}(1)^*,\underline{w}^*(2),\ldots,\underline{w}^*(n)][\underline{w}(0)^*,\underline{w}(1)^*,\ldots,\underline{w}^*(n-1]^{-1}. \tag{8}$$

This completes the proof.

Lemma 1 not only shows that $P = e^A$ is uniquely identifiable but it also gives an explicit formula. However, there are still the problems of determining A from P and of verifying the linear independence condition. These questions are not trivial in general but they are tractable in certain special cases which occur frequently in applications.

Lemma 2. Suppose A is an $n \times n$ matrix having negative and distinct eigenvalues. Let $P = e^A$. Then (i) P determines A uniquely. In fact,

$$A = \sum_{i=1}^{n} \ell n\lambda_i \, Q_i \tag{9}$$

where λ_i are the eigenvalues of P and

$$Q_i = \sum_{j \neq i} (P - \lambda_j I)/(\lambda_i - \lambda_j) \tag{10}$$

are the corresponding component matrices. Morever (ii) for any vector $\underline{c}$, the n iterates $\underline{c}\, P^k$, $k = 0,1,2,\ldots,n-1$, of P are linearly independent if and only if the n iterates of $\underline{c}$, $\underline{c}\, A^k$, $k = 0,1,2,\ldots, n-1$ are linearly independent.

Proof: (i) If the eigenvalues of A, $\nu_1,\nu_2,\ldots,\nu_n$ are negative and distinct then the eigenvalues of P, $\lambda_i = e^{\nu i}$ are positive and distinct. It is known in this case that P has at most one logarithm which is given by (9)-(10)([9,10]).

(ii) Since the eigenvalues of P are distinct, the corresponding eigenvectors $\underline{e}_i$ are complete. Thus we may write

$$\underline{c} = \alpha_1\underline{e}_1 + \alpha_2\underline{e}_2 + \dots + \alpha_n\underline{e}_n$$

and, for each positive integer k,

$$\underline{c}\, P^k = \alpha_1\lambda_1{}^k\underline{e}_1 + \alpha_2\lambda_2{}^k\underline{e}_2 + \dots + \alpha_n\lambda_n{}^k\underline{e}_n \; .$$

It is not difficult to see that the vectors $\underline{c}\, P^k$ are linearly dependent if and only if there is a polynomial $q(\lambda)$ of degree $\leq n - 1$ such that $\underline{c}\, q(P) = \underline{0}$. Moreover this condition is satisfied if and only if

$$\alpha_i q_i(\lambda_i) = 0, \quad i = 1,2,\dots,n. \tag{11}$$

Since $q(\lambda)$ had degree $\leq n - 1$ and since the eigenvalues λ_i are distinct, at least one α_1 must be zero in order for condition (11) to hold. On the other hand, if one $\alpha_i = 0$, say $\alpha_n = 0$, then choosing $q(\lambda) = (\lambda - \lambda_1)\dots(\lambda - \lambda_{n-1})$ satisfies conditions (11). Thus, the $\underline{c}\, P^k$ are linearly dependent if and only if at least one α_i is zero. However, the eigenvectors of A are identical with those of P and the eigenvalues of A are also distinct. Thus, the $\underline{c}\, A^k$ are linearly dependent if and only if at least one $\alpha_i = 0$, i.e. the $\underline{c}\, A^k$ are linearly dependent if and only if the $\underline{c}\, P^k$ are linearly dependent.
This completes the proof.

4. Application to Illness-Death Processes. The condition in Lemma 2 concerning the spectrum of A may seem restrictive, however, it suffices for the present application if the illness states are ordered according to severity of illness.

That is, an individual in state i may leave that state in only one of three ways: he enters state $i - 1$ if his condition improves, he enters state $i + 1$ if his condition worsens, or he exits the system if he dies or is lossed to follow-up. We refer to this structure as a <u>catenary illness-death process</u>. In this case A is a strictly diagonally dominant, tridiagonal matrix in which the off-diagonal entries are positive and the diagonal elements are negative. It is well known that the eigenvalues of such matrices are distinct and negative ([11-12]).

The condition that $\underline{c}\, A^k$, $k = 0,1,\ldots,n - 1$, are linearly independent is equivalent to complete observability as noted earlier.

Combining Lemmas 1 and 2 with the above remarks we obtain the following result.

<u>Theorem 3</u>. The n^2 discrete observations $y_i(t)$, $i,t = 1,2,\ldots,n$, (given by (3)) uniquely identify the intensities a_{ij} in a catenary illness-death process, provided the initial population in each of the illness states is nonempty (i.e., $\underline{m} > 0$) and the system is completely observable (i.e., $\underline{c}\, A^k$, $k = 0,1,\ldots,n - 1$ are linearly independent).

In the case $n = 2$, the matrix A is obviously tridiagonal. Thus, Theorem 3 automatically applies in this case. The complete observability condition, for $n = 2$, amounts to $a_{01} \neq a_{02}$, where $a_{0i} = -a_{1i} - a_{2i}$ are the death intensities. It is not difficult to see that if $a_{01} = a_{02}$ then

$y_i(t) = m_i e^{-a_{0i}t}$, $i = 1,2$. Thus at most $a_{01}(a_{02})$ is identifiable no matter how many values $y_i(t)$ are observed.

Turning now to the other condition, $\underline{m} > 0$, suppose either m_1 or m_2 is zero. Then (still in the n = 2 case) $y(t) = x_1(t) + x_2(t)$ is the only independent observation. Suppose $y(t)$ is known for all $t \geq 0$, i.e., we know the transfer function (Laplace transform) $\hat{\underline{y}}(s) = \underline{c}[s\ I - A]^{-1}\ \underline{m}$ $= [(m_1 + m_2)s + b_1]/[s^2 + b_2 s + b_3]$, where the b_i are certain nonlinear functions of the a_{ij}. Thus, knowing $y(t)$, $0 < t < \infty$, is equivalent to knowing the b_i. However the three b_i do not suffice to determine the four independent a_{ij}. Thus, if $\underline{m}$ is not strictly positive the system is not identifiable even if $y(t)$ is observed at all times.

Hence, in the $n = 2$ case the two conditions; $\underline{m} > 0$ and complete observability, are necessary as well as sufficient for identifiability. The above observations are stated more explicitly below.

Corollary 1. Let A be the matrix of intensities from illness states to illness states for a two-illness process. Then the four intensities a_{ij}, $i,j = 1,2$, are uniquely determined by the four discrete observations $y_i(t)$, $i,t = 1,2$, if and only if the illness states are nonvacuous at $t = 0$ and the death intensities are distinct. If either of these two conditions is violated then A is not identifiable from $y_i(t)$, $0 \leq t \leq \infty$, $i = 1,2$.

Finally, we note that when the identifiability conditions are satisfied then, in all cases $(n \geq 1)$, formulas 8-10 determine A explicitly.

REFERENCES

[1] Grewal, M. S., and K. Glover, *Identifiability of linear and nonlinear dynamical systems*, IEEE Trans. Automatic Contr., AC 21, pp. 833-837, 1978.

[2] Chiang, C. L., *Survival and stages of disease*, Math. Biosci., 43, pp. 159-171, 1979,

[3] Sacks, S. T., and C. L. Chiang, *A transition-probability model for the study of chronic diseases*, Math. Biosci., 34, pp. 325-346, 1977.

[4] Chiang, C. L., "Introduction to Stochastic Processes in Biostatistics," Wiley, New York, 1968.

[5] Fix, E., and J. Neyman, *A simple stochastic model of recovery, relapse, death and loss of patients*, Hum. Biology, 23, pp. 205-241, 1951.

[6] Gazes, P. C., E. N. Mobley, Jr., H. M. Faris, Jr., R. C. Duncan, and G. B. Humphries, *Preinfarctional (unstable) angina-a prospective study-ten year follow-up*, Circulation, 48, pp. 331-337, 1970.

[7] Bellman, R., and K. J. Astrom, *On structural identifiability*, Math. Biosci., 7, pp. 329-339, 1970.

[8] Zadeh, L. A., and C. A. Desoer, "Linear System Theory: The State Space Approach," McGraw-Hill, New York, 1963, pp. 502.

[9] Singer, B., and S. Spilerman, *The representation of social processes by Markov models*, Amer. J. Sociology, 82, pp. 1-54, 1976.

[10] Gantmacher, F. R., "The Theory of Matricies," Vol. 1, Chelsea, New York, 1960, pp. 239.

[11] Sandberg, S., D. H. Anderson, and J. Eisenfeld, *On Identification of compartmental systems*, "Applied Nonlinear Analysis," V. Lakshmikantham (ed.), Academic Press, New York, pp. 531-542, 1979.

[12] Ortega, J. M., "Numerical Analysis," Academic Press, New York, pp. 106.

COMPUTER SYMBOLIC SOLUTION OF NONLINEAR ORDINARY DIFFERENTIAL EQUATIONS WITH ARBITRARY BOUNDARY CONDITIONS BY THE TAYLOR SERIES

James N. Hanson

Cleveland State University

INTRODUCTION

Many theoretical tools are available for obtaining qualitative properties of the solutions of differential equations, such as solution existence over an interval of the independent variable, periodicity, quasi-periodicity, boundedness, etc. However, these same tools seldom have been used in actually exhibiting the analytic solution or an analytic approximation of the solution. Of these many tools, the author (1,2) has examined the automatic analytic capabilities for obtaining efficient solutions by the method of successive approximations, the method of steepest descent, Newton's method and by expansion into a Taylor's series (1 - 3). Numerical application of these methods is well known (4,5). However, the author's purpose here and elsewhere has been to apply these methods in a purely analytical manner irrespective of the enormous amount of algebraic and symbolic manipulation required in the intermediate steps leading to the solution.

A typical operation required in using Newton's method to solve $x'' = f(x',x,t)$ might be the construction of the

ISBN 0-12-045550-1

antiderivative of

$$e^{at}\, t^{\ell} \sin^{m}(bt) \cos^{n}(ct)$$

The antiderivative of this expression is readily formulated as a recursion formula, however, even for small ℓ, m or n manual explicit expansion becomes a hopeless task. This already immense task is further compounded algebraically and must in turn be integrated, and so forth. The computer can perform these manipulations readily.

The benefit of polynomial manipulation is that the size of expressions do not grow so rapidly. It is the generation of power series, i.e., the Taylor expansion, that is the subject of this paper.

Computer Algebraic and Symbolic Manipulation. The analytical mathematical use of the digital computer, is well established. There have been a number of papers dealing with the automatic computer generated analytic solution of ordinary differential equations (7 - 9). Formac-PL/I (10) has been used due to its general availability and compatibility, with small main storage, even as little as 200K bytes. All the examples in this paper were run in a partition of less than 300K bytes.

The benefits of symbolic methods are the possible attainment of very high accuracy, and the compactness and tractability of an analytic expression. In some cases just expressing the differential equation is an impossible manual task. For example, Hanson (11) has used symbol manipulation in order to express the differential equations for integration by the Runge-Kutta procedure, for sliding cam motion.

Solution by Taylor Series. The Taylor series solution of a differential equation may be obtained various ways. For example, $x' - \tanh t \sec x = 0$, $x(0) = 0$ has the exact solution $x^*(t) = \sin^{-1}(\ln \cosh t)$, $0 \le t \le \cosh^{-1} e$ where the Taylor series (1) expanded about $t = 0$ for $x^*(t)$ may be obtained by successive approximations (Piccard's method) by replacing $\tanh t$ and $\sec x$ by their MacClaurin expansions and starting from $x(t) = 0$ or from Newton's method. A simpler procedure is to construct the Taylor series solution by recursively generating higher order derivatives from the differential equation just as it is found. The method lies in comparitive disuse due to the enormity of the intermediate algebraic manipulation. However, it has to its credit great simplicity and straightforwardness. Specifically, very complicated nonlinear equations with arbitrary multi-point boundary values can be handled.

It will be assumed that $x(t)$, satisfying $f(t,x,x',\ldots x^{(n)}) = 0$ and subject to n distinct boundary conditions $x^{(m_i)}(t_i) = c_i$, possesses continuous derivatives up to some order N sufficiently larger than n and the largest m_i within some interval of interest along the t-axis which contains the t_i and the point of expansion. It will be further assumed that the highest derivative can be isolated. Consider the TPBVP, $x'' = e^x$, $x(0) = x(1) = 0$, to be expanded about $c = 0$ and define the missing derivative by $x'(c) = x'(0) = z$. Successive recursive differentiations of $x' = e^x$ provides

$$
\begin{aligned}
x(c) &= 0 \\
x'(c) &= z \\
x''(c) &= (e^x)_{t=0} = 1 \\
x'''(c) &= (e^x x')_{t=0} = z \\
x^{(4)}(c) &= (e^x (x')^2 + e^x x'')_{t=0} = (e^x (x')^2 + (e^x)^2)_{t=0} \\
&= z^2 + 1 \\
x^{(5)}(c) &= \ldots = z^3 + 3z + z \\
&\ldots
\end{aligned}
$$

The resulting identity for z for the fifth order approximation is

$$0 = x(1) = 0 + z \cdot 1 + 1 \cdot 1^2/2! + z \cdot 1^3/3! + (z^2 + 1) \cdot 1^4/4! + (z^3 + 3z + z) \cdot 1^5/5!$$

Table I shows results for orders of approximation up to 10. If $c = .5$ had been chosen, then two identities (polynomials) in $x(c)$ and $x'(c)$ would have resulted. The order of approximation will be indicated by NPT (number of polynomial terms).

TABLE I. Solution of $x'' = \exp(x)$, $x(0) = x(1) = 0$ versus order of approximation (NPT). The last row gives the exact solution.

NPT	Z(I)	x(0.2)	x(0.4)	x(0.6)	x(0.8)
3	-0.428571	-0.066286	-0.096000	-0.092571	-0.059429
4	-0.472251	-0.074998	-0.112633	-0.113747	-0.077226
5	-0.458005	-0.072136	-0.106962	-0.106008	-0.070106
6	-0.465662	-0.073677	-0.110065	-0.110445	-0.074518
7	-0.462506	-0.073042	-0.108777	-0.108549	-0.072509
8	-0.464116	-0.073366	-0.109436	-0.109538	-0.073613
9	-0.463381	-0.073218	-0.109135	-0.109080	-0.073080
10	-0.463750	-0.073292	-0.109286	-0.109312	-0.073359
	-0.463633	-0.073268	-0.109238	-0.109238	-0.073268

This procedure is easily generalized for any equation of the form $x^{(n)} = f(t,x,x',\ldots,x^{(n+1)})$ and any set of boundary values.

The precision of this procedure is governed by the mode of numerical constants occuring in the differential equation and in the initial conditions. Numerical constants may be of two forms, floating point or rational, e.g., 2.3 is considered to be in floating point form whereas 23/10 is its rational counterpart. The advantage of structuring a problem in rational form is that rational arithmetic is performed thus permitting almost arbitrarily high precision. Hanson and Russo [19] have adapted Formac so that rational arithemtic will not result in such large (in digit length) numbers that storage overflow occurs, and so that rational and floating point arithmetic may be mixed in order to maintain accuracy with the former and benefit from the computational speed of the latter. When more than one missing derivative must be solved for by Newton's method, Guassian elimination has been used.

The Computer Program. A Formac-PL/I computer program for implementing the Taylor series solution of $x^{(n)} = f$ includes the following steps:

(1) input,

(2) construction of the Taylor series of $x(t)$ in terms of missing derivatives,

(3) application of boundary values to yield an algebraic system of equations in the missing derivatives,

(4) solution for missing derivatives by Newton's method using Gaussian elimination,

(5) missing derivatives substituted into the Taylor series,

(6) output.

The solution may be continued by using the terminal values from one interval as the initial values of the next. The differential equation $x(n) = f$ is input as a string in symbols D(0), D(1),...,D(N - 1), e.g., $x'' = 3x + x' \sin t$ is entered as F = '3*D(0) + D(1)* SIN(T)'.

No simple means is available for estimating the missing derivatives. However, as with the case of $x'' = \exp(x)$, a lower order solution may be manually tractable, or the program may be used in a trial and error fashion.

As an example, consider the following complicated multipoint boundary value

$$
\begin{aligned}
x(4) &- tx''' + x'x'' + (x')^2 - \sin t \cdot x + \exp(x') \\
&= 1/2 \exp(\sin t)(1/2 \sin^2 2t - 6\cos 2t - 2\sin t) \\
&- 1/2\, t \exp(\sin t) \cdot \sin 2t \cdot (\sin t - 3) \\
&+ (\exp(\sin t) \cdot \cos t + \pi^{-1})\exp(\sin t) \cdot (\cos^2 t - \sin t) \\
&+ (\exp(\sin t) \cdot \cos t + \pi^{-1})^2 \\
&- \sin t \cdot (\exp(\sin t) + \pi^{-1} t) \\
&+ \exp(\exp(\sin t) \cdot \cos t + \pi^{-1})
\end{aligned}
$$

for $x(\pi) = 2$, $x'(\pi) = -1 + \pi^{-1}$, $x(2\pi) = 3$ and $x'(2\pi) = 1 + \pi^{-1}$ and where $c = 3\pi/2$, NPT = 6 with estimates $x(c) = 2$, $x'(c) = 1/3$, $x''(c) = 1/3$ and $x'''(c) = 1/10$. The computer output is shown in Table II. The solution of this problem is $x(t) = t\pi^{-1} + \exp(\sin t)$. The limiting factor in solving such problems is the large storage needed to construct high order derivatives of f. This difficulty is avoided by using small NPT over many continued relatively small intervals, thus incurring the advantages of spline representations. Epsilon specifies an upper bound on the accuracy of the Newton's method solution for the missing derivatives, and, NORM is the relative Euclidian distance between two successive approximations.

TABLE II. Computer output for complex fourth order equation.

I	T	APPROX SOL	EXACT SOL	DIFFERENCE	APPROX DER
0	0.00000	-2.14175E+00	1.00000E+00	3.14175E+00	1.72847E+01
1	0.47124	3.30038E+00	1.72458E+00	1.57580E+00	6.77636E+00
2	0.94248	4.99888E+00	2.54570E+00	2.45318E+00	1.08341E+00
3	1.41372	4.81460E+00	3.13502E+00	1.67958E+00	-1.45698E+00
4	1.88496	3.92035E+00	3.18844E+00	7.31907E-01	-2.11021E+00
5	2.35619	2.97647E+00	2.77811E+00	1.98354E-01	-1.12829E+00
6	2.82743	2.28302E+00	2.26209E+00	2.09386E-02	-1.12829E+00
7	3.29867	1.90880E+00	1.90519E+00	3.61044E-03	-4.83004E-01
8	3.76991	1.79692E+00	1.75556E+00	4.13633E-02	-2.93309E-02
9	4.24115	1.84724E+00	1.76024E+00	8.70007E-02	2.11814E-01
10	4.71239	1.97542E+00	1.86788E+00	1.07542E-01	3.20159E-01
11	5.18363	2.14869E+00	2.06024E+00	8.84436E-02	4.26657E-01
12	5.65487	2.39835E+00	2.35556E+00	4.27908E-02	6.64057E-01
13	6.12611	2.80899E+00	2.80519E+00	3.79916E-03	1.11748E+00
14	6.59734	3.48437E+00	3.46209E+00	2.22880E-02	1.77499E+00
15	7.06858	4.49008E+00	4.27811E+00	2.11965E-01	2.47818E+00
16	7.53982	5.77281E+00	4.98844E+00	7.84366E-01	2.87273E+00
17	8.01106	7.05643E+00	5.23502E+00	1.82141E+00	2.35898E+00
18	8.48230	7.71473E+00	4.94570E+00	2.76903E+00	4.25208E+00
19	8.95354	6.62084E+00	4.42458E+00	2.19626E+00	-5.31523E+00
20	9.42478	1.97345E+00	4.00000E+00	2.02655E+00	-1.53465E+01

THE DIFF EQ,APPROX & EXACT SOL ARE:

$$D(4)=-1/2\ \#E^{SIN\ (T)}\ SIN(2T)\ (\ SIN(T)-3)\ T + X^{(3)}.(T) + SIN$$

$$(T)\ X.(T) - SIN\ (T)\ (T/\#P + \#E^{SIN\ (T)} + 1/2\ \#E^{SIN\ (T)}$$

$$(1/2\ SIN^2\ (2T) - 2\ SIN\ (T) - 6\ COS\ (2T)) + \#E^{SIN\ (T)}\ (COS^2\ (T)$$

$$-\ SIN\ (T))(\#E^{SIN\ (T)}\ COS\ (T) + 1/\#P) - X^{(2)}.(T)\ X^{(1)}\ .(T) +$$

$$(\#E^{SIN\ (T)}\ COS\ (T) + 1/\#P)^2 - X(1)\ .^2\ (T) + \#E^{\#E^{SIN\ (T)}}\ COS$$

$$(T) + 1/\#P - \#E^{X(1)}\ .(T)$$

SERIES = 17.2846827 T - 14.4082335 T^{2} + 5.16618508 T^{3} - .9337219 T^{4} + .08382693 T^{5} - .00295403 T^{6} - 2.14174788

SOLUTION = T/#P + #E$^{\text{SIN (T)}}$

PT OF EXPANSION,NEWT METH ACC & NORM:

CC=3/2 #P

EPSILON = .99999999E-05

NORM = .6394398E-05

SUMMARY OF BOUNDARY CONDITIONS:

D**0(X(#P))=2

D**1(X(#P))=-1+1/#P

D**0(X(2*#P))=3

D**1(X(2*EP))=1+1/#P

SUMMARY OF INIT EST OF MISSING DERIVATIVES:

D**0(X(C))=?=2

D**1(X(C))=?=1/3

D**2(X(C))=?=1/3

D**3(X(C))=?=1/10

Thermionic currents. The important application of thermionic currents supply a good test for the method of this paper. A typical equation (12) is $x'' - 2 \sinh x + u(t) = 0$ with boundary values $x''(-1) = x''(1) = 0$ where $u(t) = 0$ $(t \le 0)$ and $u(t) = b(t > 0)$. The solution for $a = 30$ and $b = -10$ for $c = 0$ and estimates $x\ (0) = 10$, $x'(c) = -1$, and setting NPT = 5 is:

$$x(t) = -8.34476215t - 35.1502239t^2 + 42.4977629t^3 + 5.85837066t^4 - 12.7493288t^5 + 5.34855982$$

The same problem solved for the equivalent boundary values $x(-1) = \sinh^{-1}(a/2)$ and $x(1) = \sinh^{-1}(b/2)$ gives

$$x(t) = -8.2300795t - 35.6156822t^2 + 39.7548086t^3 + 30.80766399t^4 - 34.3821016t^5 + 5.35297645$$

These two solutions agree numerically to sufficient accuracy and mutually converge for increasing NPT.

The Two Body Problem. The two body problem presents a useful test for the trade off of truncation error of the Taylor series versus the round-off error propogated by continuation. The two body problem is

$$r'' - rv'^2 + ar^{-2} = 0, \quad r^2v' = h$$

where a is the field strength, h the angular momentum, r is the radial distance from the central body and v is the true anomaly. If v is eliminated an equation in r results.

$$r'' = (1 - e^2)\, r^{-3} - r^{-2}$$

where a unit semi-major axis and an orbital period of 2π have been chosen. Further let the boundary values be at the pericenter at time $t = 0$, $r(0) = a(1 - e) = 1 - e$, $r'(0) = 0$ where e is the eccentricity. Alternately, under the same conventions, r can be eliminated to give an equation in v,

$$v''' = (3/2)v'^{-1}v''^2 - 2v'^3 + 2(1 - e^2)^{-3/4}\, v'^{5/2}$$

with initial conditions $v'(0) = (1 + e)^{1/2}\,(1 - e)^{-3/2}$, $v(0) = v''(0) = 0$. These two equations were solved for various NPT and number of continuations. The relative absolute error at one revolution for NPT = 5 and 20 continuations is 10^{-5} for both r and v. Each increase in NPT increases the accuracy by about one digit, however, increasing the number of continuations reduces the accuracy. Similar results were obtained for the first three derivatives.

Special Functions. Special functions can be defined as the solution of a differential equation. Automatic Taylor series solution can be used to produce the exact expansion to as many terms as desired. This followed by evaluation to very high precision can provide special function values to 50 places, or greater, if needed.

For example the exponential function, exp(t), can be expanded by solving $x' - x = 0$, $x(0) = 1$ for $c = 0$. Similarly, Bessel functions, Mathieu functions, Weierstrass functions, or un-named functions can be represented and tablized. Another application arises in the evaluation of definite integrals. The integral of $f(u)$ between t and $u = a$ can be expressed as a power series in t by solving the equivalent differential equation, $x' = f(t)$, $x(a) = 0$.

Variational Calculus. This brute force method is especially useful in actually obtaining solutions to problems arising from variational principles. If one invokes the variational calculus of Pontryagin's maximum principle, then very complicated differential equations usually result. In such cases a straight-forward Taylor series solution is particularly welcome to the engineer or physicist.

For example Fermats principle for geometric optics requires a light ray to follow a path such that the integral of the index of refraction, n, is minimized,

$$\min_{x(t)} \int_{s_0}^{s_1} n(x(t),t)\, ds$$

which upon applying Euler's equation gives

$$x'' = (1 + x'^2)((1 + x'^2)n_x/n - x'n_t n)$$

where subscripts indicate partial differentiation.

As an indication of the accuracy obtainable, the Euler equation for the caterary problem is

$$x'' = (1 + x'^2)^{1/2} , \quad x(0) = 1, \quad x'(0) = 0$$

was solved for a single 10th order polynomial in the range $t = 0$ to $t = 1$, this polynomial solution giving 8 place accuracy when compared to the exact solution, $x(t) = \cosh t$.

Fluid Mechanics. The behavior of a fluid boundary in the shear range as formulated by the Dukler [13] provides a very complicated equation, but one which must, nevertheless, be solved. Dukler's equation is

$$x''' = (-xx'' + a(1 - x') - 2btx''^2 (1 - \exp(-ct))$$

$$(1 - ct \exp(-ct))) \quad (1 + 2bt^2 x''(1 - \exp(ct)))^{-1}.$$

The solution for $x(t)$ gives the velocity profile x in terms of distance t. The boundary values for this problem are $x(0) = 0$, $x'(0) = 0$ and $x'(\infty) = 1$ where an infinite distance from the boundary may be approximated by $t = 1$. Our computer program provided excellent agreement with Roberts' and Shipman's numerical solution [13].

Test Problems for Numerical Methods. A large literature exists for testing various difficult differential equations. Test problems usually present some difficulty such as stiffness, rapid variations or implicitness (i.e., the highest derivative not being conveniently separated algebraically). These difficulties inhibit the accuracy of finite difference methods [14-16].

The Taylor series method have proved particularly useful on implicit differential equation. For example, consider Wolfe's test problems [15],

$$f(x',x,t) = t^2x'^5 + x' - tx - 1 = 0, x(0) = 0$$

$$f(x',x,t) = x'^5 - x' + x - e^{5+} = 0, \ x(0) = 1$$

whose exact solutions are $x(t) = t$ and $x(t) = \exp(t)$, respectively. The exact solutions were obtained in both cases. These problems were also solved by converting them to explicit problems by differention.

Many test problems from Aziz [14], Roberts [13], and Cash [16] have been accurately solved. For example, Aziz poses

$$x'' - 400x = 400 \cos^2 \pi t + 2\pi^2 \cos 2\pi t$$

$$x(0) = 0, \ x(1) = 0$$

whose exact solution is

$$x(t) = e^{-20}e^{20t}(1 + e^{-20})^{-1} + e^{-20t}(1 + e^{-20})^{-1} - \cos^2 \pi t.$$

The Taylor series with rational coefficient was obtained by expanding about $c = 0$.

The Pendulum. The pendulum presents an example of a complex differential equation whose solution is required to great accuracy, e.g., in chronometry and gravimetry. The gravimetric use of the torsion pendulum requires the solution of

$$x'' = a(b_1 + c_1(x - x(0))^2)^4 \ (b_2 + c_2(x - x(0))^2)^{-1} \ x'$$

where $x(t)$ is the angular displacement whose value at $t = 0$ is $x(0)$. The complexity of the torsion pendulum arises due to the lengthening of the string as x changes (17). Since the solution, specifically the time, is needed only from $x = (0)$ to $x = 0$, a relatively low order Taylor series (NPT = 10) yields accuracy to 6 to 7 places.

Miscellaneous Problems. The following problems from Davis [18] have been accurately solved.

Equation	Name, Application
$x'' = (x - 1)(2 - x)/x^5 = 0$	Electron motion in geomagnetic field
$x'' + 2x'/t + f(t) = 0$	Emden, self gravitating gas
$x'' + ax + bx^3 + c \sin kt = 0$	Duffing, nonlinear vibration
$x'' - a(1 - x^2)x' + bx'3 = 0$	Stablizing ship's roll
$xx'' - x'^2 - (x^2 - x)x' + 2(x^2 - x^3) = 0$	Volterra, hysteresis Phenomenon
$x'' + x + a + bx^2 = 0$	Relativistic pericentric precession
$(1 + ax)x'' + x + b + cx'^2 = 0$	Gerber's retarded potential pericentric precession
$x''' + xx'' + a(1 - x'^2) = 0$	Falkner-Skan, boundary layer flow
$x'' - t^{-1/2}\ x3/2 = 0$	Thomas-Fermi
$3xx'' + x'^2 + 4xx' + x^2 = 0$	Langmuir, ionized gas dynamics

Higher order equations have been tested, e.g., the complicated rubber band pendulum [18]

$$(x')^2x'''' + (x'x'')x''' + ((a - 2b)(x')^2 - 2c(x')^4 - ax'x'' + 2axx'')\ x'' + (-2abxx')x' = 0$$

Such equations arise and must be dealt with.

REFERENCES

[1] Hanson, J. N., *Experiments with equation solutions by functional analysis algorithms and formula manipulation*, J. Computational Physics, 9,26-52, Feb. 1972.

[2] Hanson, J. N., *Functional analysis, formula manipulation, and satellite geodesy*, J. Geophysical Research, 78, pp. 3260-3270, June, 1973.

[3] Kantorovich, L. V., and V. I. Krylov, "Approximate Methods of Higher Analysis", P. Noordhoff, Groningen, Netherlands, 1964.

[4] Antosiewicz, H. A., and W. C. Rheinbolt, "Numerical Analysis and Functional Analysis", (J. Todd Ed.), McGraw-Hill, New York, 1962.

[5] Collatz, L., "Functional Analysis and Numerical Mathematics", Academic Press, New York, 1966.

[6] *Special interest group in symbol and algebraic manipulation (SIGSAM) bulletin*, published by the Association for Computing Machinery, New York.

[7] Kjaer, J., "Computer Methods in Solution of Differential Equations", Haldor Topsoe, Vedbaek, Denmark, 1972.

[8] Norman, A. C., *Expanding the solutions of implicit sets of ordinary differential equations in power series*, The Computer Journal, 19, pp. 63-68, 1976.

[9] Barton, D., I.M. Willers, and R.V.M. Zahar, *Taylor series methods of ordinary differential equations - an evaluation*, "Mathematical Software", J. Rice Ed., pp. 369-390.

[10] Xenakis, J., "Pl/I Formac Interpreter User's Reference Manual", IBM, 1967.

[11] Hanson, J.N., *Planar motion of sliding cams by computer algebraic manipulation, mechanisms and machine theory*, 14, pp. 111-120, 1979.

[12] Davis, J.T., "Introduction to Nonlinear Differential and Integral Equations", Dover Publications, New York, 1962.

[13] Roberts, S. M., "Two-Point Boundary Value Problems: Shooting Methods", American Elsevier Publishing Company, New York, 1977.

[14] Aziz, A. K., (Ed.), "Numerical Solution of Boundary Value Problems for Ordinary Differential Equations", Academic Press, New York, 1975.

[15] Wolfe, M. A., *The numerical solution of implicit first order ordinary differential equations with initial conditions*, The Computer Journal, 14, pp. 173-178, May 1971.

[16] Cash, J. R., *Semi-implicit Runge-Kutta procedures with error estimates for numerical integration of still systems of ordinary differential equations*, J. ACM, 23, pp. 445-460, July 1976.

[17] Allen, M., and E. J. Saxl, *Elastic torsion in wires under tension*, J. Applies Physics, 40, 2505-2509, May 1969.

[18] Allen L. Kirg, *Oscillations of a loaded rubber band*, Am. J. of Physics, 42, 699-70 , August 1974.

[19] Hanson, J. N., and P. Russo, *Some data conversions for managing the internal and output form of formac constants*, ACM-SIGSAM Bulletin, 10, 21-26, May 1976.

A NOTE ON NONCONTINUABLE SOLUTIONS OF A DELAY DIFFERENTIAL EQUATION

T. L. Herdman

Virginia Polytechnic Institute
and State University

INTRODUCTION

In this note we discuss the behavior of certain noncontinuable solutions of the n-dimensional retarded differential equation

$$x'(t) = q(t, x(t - \tau(t))), \ 0 \leq t < T, \tag{1.1}$$

with initial data

$$x(t) = \varphi(t), \ t \in [-r,0], \tag{1.2}$$

where $\varphi : [-r,0] \to \mathbb{R}^n$ is continuous function, r and T are positive constants, $q : [0,\infty) \times \mathbb{R}^n \to \mathbb{R}^n$ and $\tau : [0,+\infty) \to [0,+\infty)$ are continuous functions. In particular, we investigate the behavior of $\lim|x(t)|$ as $t \to T^-$ given that $x(t)$ is a noncontinuable solution of (1.1)-(1.2) having the property that $\lim_{t \to T^-} \sup|x(t)| = +\infty$. For ordinary differential equations, that is equation (1.1) with $\tau \equiv 0$, it is well known that the continuity of the function q is sufficient to assure the property that $\lim_{t \to T^-} |x(t)| = \lim_{t \to T^-} \sup|x(t)| = +\infty$ for the noncontinuable solution $x(t)$. However, the noncontinuable solutions of (1.1)-(1.2) do not in general have this property. In fact, one can construct an

ISBN 0-12-045550-1

example of a noncontinuable solution $x(t)$ of (1.1)-(1.2) such that $|x(t)|$ is bounded on $[0,T)$, (See [2, page 21].

In section 2, we construct a counter example to the following conjecture of Burton and Grimmer [1].

Conjecture (C). Let the functions q, τ and φ satisfy the above conditions. If $x(t)$ is a noncontinuable solution of (1.1)-(1.2) then $\lim_{t \to T^-} |x(t)| = +\infty$.

II. Example. We construct here a retarded differential equation having a noncontinuable solution $x(t), t \in [-1,1)$, such that $\lim_{t \to T^-} \sup |x(t)| = +\infty$ but $\lim_{t \to T^-} |x(t)| \neq +\infty$. The function q, τ and φ of this equation will satisfy the stated conditions of section 1 and therefore the equation together with its solution shows that the conjecture (C) is false.

As a first step in the construction of the example, we define the following sequences $\{\alpha_n\}$, $\{a_n\}$, $\{b_n\}$ and $\{\beta_n\}$.

$$\alpha_o = 0; \ \alpha_n = \frac{2^n - 1}{2^n}, \ n = 1, 2, \dots . \tag{2.1}$$

$$a_o = 0; \ a_n = \alpha_n + \frac{1}{2^{n+3}}, \ n = 1, 2, \dots . \tag{2.2}$$

$$b_o = 0; \ b_n = \alpha_n + \frac{1}{2^{n+2}}, \ n = 1, 2, \dots . \tag{2.3}$$

$$\beta_o = \frac{1}{4}; \ \beta_n = \alpha_n + \frac{3}{2^{n+3}}, \ n = 1, 2, \dots . \tag{2.4}$$

From (2.1)-(2.4), we note that each of the sequences $\{\alpha_n\}$, $\{a_n\}$, $\{b_n\}$ and $\{\beta_n\}$ converge to 1 as $n \to \infty$; also it follows that

$$0 = \alpha_o = a_o = b_o < \beta_o \text{ and } \beta_{n-1} < \alpha_n < a_n < \beta_n$$
$$\text{for } n = 1, 2, \dots . \tag{2.5}$$

The function $x : [-1,1) \to \mathbb{R}$ is defined as follows:

$x(t) = 1$ for $t \in [-1,0];$ (2.6)

$x(\beta_n) = x(b_n) + (-1)^{n+1} (\frac{1}{2})^{n+1}$ for $n = 0, 1, 2, \ldots;$ (2.7)

$x(\alpha_n) = x(\beta_n)$ for $n = 1, 2, \ldots;$ (2.8)

$x(t) = (-1)^n n$ for $t \in [a_n, b_n]$, $n = 1, 2, \ldots;$ (2.9)

$$x(t) = \left(\frac{x(\alpha_n) - x(\beta_{n-1})}{\alpha_n - \beta_{n-1}}\right)(t - \alpha_n) + x(\alpha_n) \text{ for } t \in [\beta_{n-1}, \alpha_n] \qquad (2.10)$$

$(-1)^n x(t)$ is decreasing and continuously differentiable on $[b_n, \beta_n]$ for $n = 1,2,\ldots;$ (2.11)

$(-1)^{n+1} x(t)$ is decreasing and continuously differentiable on $[\alpha_n, a_n]$. (2.12)

In view of (2.6)-(2.12), $x(t)$ is a continusously differentiable function defined on $[-1,1]$ and satisfied $\limsup_{t\to 1^-} |x(t)| = +\infty$; however, $\lim_{t\to 1^-} |x(t)| \neq +\infty$. The graph $C = \{(t,x(t))\ t \in (-1,1)\}$ is shown in Figure 1.

We now define a continuous function $\tau: [0,\infty) \to [0,1]$ by, $\tau \equiv 0$ on $[1,+\infty]$.

$\tau(0) = 1;\ \tau(t) = \frac{1}{2}$ for $t \in [\beta_0, \alpha_1];$ (2.11)

$\tau(t) = b_n - \alpha_n$ for $t \in [b_n, \alpha_{n+1}]$, $n = 1,2,\ldots;$ (2.12)

$\tau(a_n) = \frac{1}{2}(\tau(\alpha_n) - \tau(b_n))$, $n = 1,2,\ldots;$ (2.13)

τ is continuous on $[\alpha_n, a_n]$ and $[a_n, b_n]$, $n = 1, 2, \ldots;$ (2.14)

$$\tau(t) = \left(\frac{\tau(\beta_0) - 1}{\beta_0}\right)t + 1 \text{ for } t \in [\alpha_0, \beta_0]. \qquad (2.15)$$

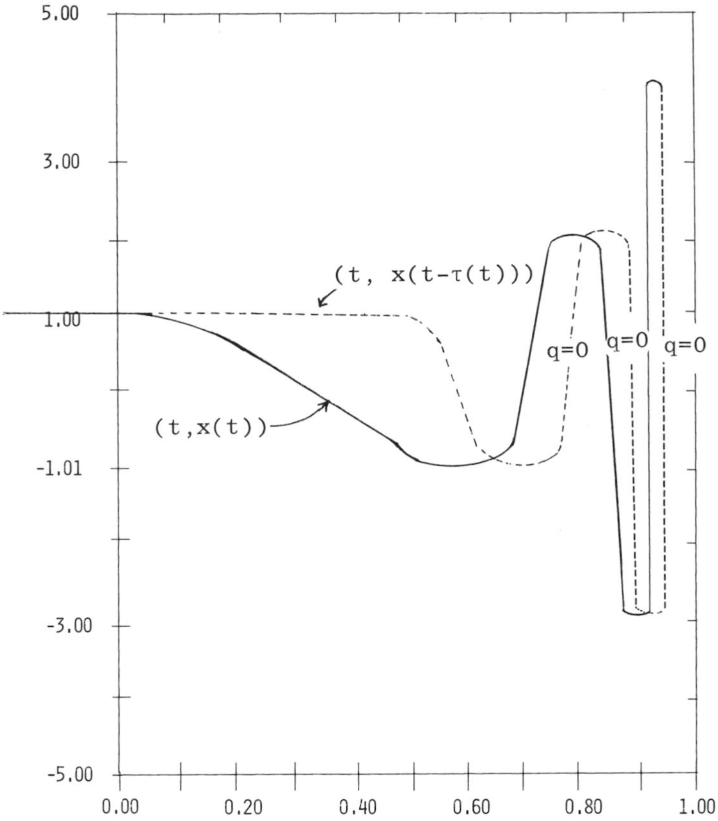

FIGURE 1

The graph $H = \{t,\tau(t)) \mid t \geq 0\}$ is shown in Figure 2. We note that the continuous function $\omega(t) = t - \tau(t)$ satisfies $\omega(\alpha_{n+1}) = b_n$, $\omega(\beta_n) = a_n$, $\omega(b_n) = \alpha_n$ and $\omega(a_{n+1}) = \beta_n$ for $n = 2, 3, \ldots$. The graph $C^* = \{(t,x(t - \tau(t)))\ t \in [0,1]\}$ is shown in Figure 1.

To complete the example, we define the function $q\colon [0,\infty) \times \mathbb{R} \to \mathbb{R}$ as follows

$$q(t,x(t - \tau(t)) = x'(t) \quad \text{for} \quad t \in [0,10); \tag{2.16}$$

$$q(1,\eta) = 0 \quad \text{for} \quad \eta \in \mathbb{R}; \tag{2.17}$$

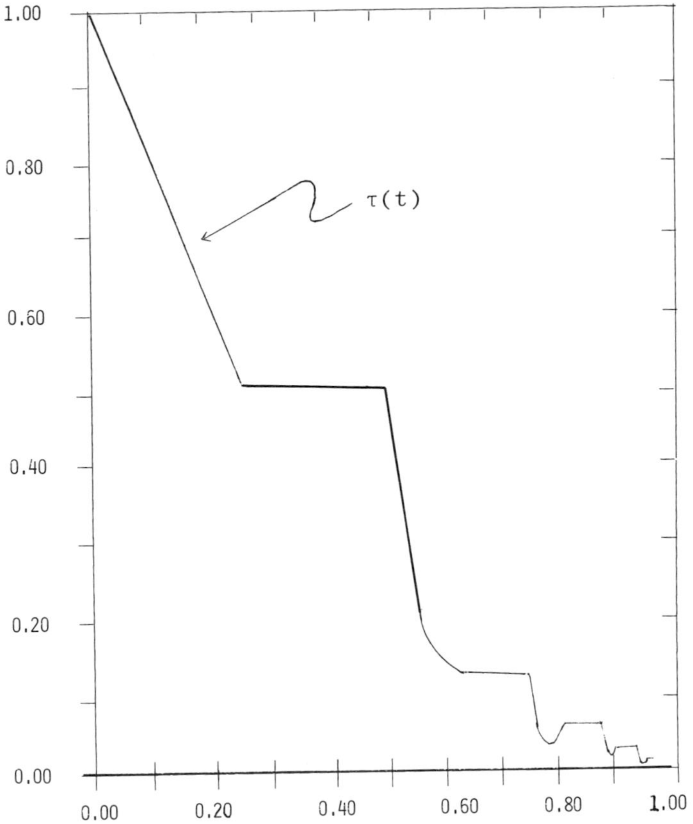

FIGURE 2

$$q \text{ is define on } ([0,\infty) \times \mathbb{R}) - (C^* \cup \{(1,n)\ n\ \mathbb{R}\}) \text{ such that } q \text{ is continuous on } [0,\infty) \times \mathbb{R}. \tag{2.18}$$

The function $x(t)$ is continuously differentiable, thus (2.16) implies that q is continuous on C^*. In view of (2.16), we have that $q(t,x(t\ \ \tau(t))) = 0$ for $t \in [a_n,b_n]$, $n - 2, 3, \ldots$. Equations (2.16) and (2.17) yield the continuity of q on $C^* \cup \{(1,n) \mid n \in \mathbb{R}\}$ a closed subset of $[0,\infty) \times \mathbb{R}$. This allows the continuous extension (2.18) of q.

It now follows that $x(t)$ is a solution of the system (1.1)-(1.2) where $\varphi(t) \equiv 1$ on $[-1,0]$. Consequently, the solution x provides a counterexample to conjecture (C).

REFERENCES

[1] Burton, T. and R. Grimmer, *Oscillation, continuation and uniqueness of solutions of retarded differential equations*, Trans. Amer. Math. Soc. 187 (1974).

[2] Myskis, A. D., *General theory of differential equations with a retarded argument*, Uspehi Mat. Nauk, Vol. 4, 1949, No. 5 (23), pp. 99-41 (Ressian). (Available in English as Am. Math. Soc. Trans., No. 55, 1951.)

THE CENTER OF A FLOW

Ronald A. Knight
Northeast Missouri State University

Birkhoff directed considerable attention toward developing a theory for qualitatively determining all types of solutions or motions and their interrelationships for dynamical systems. He introduced the classical concept of central motions in his paper, *Uber gewisse zentrale Bewegungen dynamischer Systeme*, [3] in 1926. Certain notions from his earlier work were integrated with the central motion concept in his book, "Dynamical Systems," [4] orginally published in 1927.

Birkhoff referred to antonomous systems of differential equations of the general form $dx_i/dt = f_i(x_1,...,x_n)$, $i = 1, ..., n$ having right members continuous in some region of R^n as dynamical systems. He demonstrated that for a compact n-dimensional manifold M there is a set M_1 of central motions (nonwandering motions) towards which all other motions of the system tend asymptotically. Using transfinite induction, he obtained the greatest closed subset M_0 all of whose points are nonwandering with respect to M_0. Birkhoff called this collection the set of central motions of M and observed that it coincides with the closure of the set of Poisson stable points.

ISBN 0-12-045550-1

In their treatise, "Qualitative Theory of Differential Equations," [6] Nemytskii and Stepanov use Birkhoff's construction in generalized dynamical systems on compact metric phase spaces obtaining identical results. The extension to a locally compact or complete metric phase space is a straight forward next step as Bhatia and Hajek note in their monograph, *Theory of dynamical systems*, [1]. In view of these results, the set of central motions or center of a continuous flow on a Hausdorff phase space has been defined to be the closure of the set of Poisson stable points (see [1]).

In the paper, *Central motions*, [5] Knight generalized the result that the set of Poisson stable points is dense in the phase space to locally compact Hausdorff spaces. Our task here is to show that this statement is true for completely regular (uniform) Baire Hausdorff phase spaces.

Throughout the following, we assume that there is a given flow (X,π) on a Hausdorff phase space X. We shall denote the orbit, limit set, and prolongational limit set of x in X by $C(x)$, $L(x)$, and $J(x)$, respectively. The unilateral sets carry the appropriate $+$ or $-$ superscript. A point x is *positively (negatively) Poisson stable* provided $x \in L^+(x)$ ($x \in L^-(x)$) and x is *Poisson stable* if it is both positively and negatively Poisson stable. A point x is *nonwandering* if $x \in J(x)$. The reader may consult [1] and [2] for basic dynamical system concepts used herein.

We now obtain our major result.

Theorem: Let (X,π) be a nonwandering flow on a completely regular Baire space X. Then the set of Poisson stable points is a dense G_δ set of second category in X.

Proof: A completely regular Hausdorff space is uniformizable and every uniform structure can be determined by a set of pseudometrics. Let G be the gage of an admissible uniformity for the topology on X. We now construct a sequence of sets (F_n) similar to those constructed in Theorem 5.10 of [6]. For each positive integer n define $F_n^+ = \{x \in X : \rho(x,xt) \geq 1/n$ for $t \geq n$ and $\rho \in G\}$ and $F_n^- = \{x \in X : \rho(x,xt) \geq 1/n$ for $t \leq -n$ and $\rho \in G\}$. Let $F_n = F_n^+ \cup F_n^-$ for each n.

First, we shall show that F_n is closed. Let (x_i) be a net in F_n^+ where $x_i \to x$. For each $\rho \in G$ and $t \geq n$ we have $\rho(x_i,x_it) \geq 1/n$ and so $\rho(x_i,x_it) \to \rho(x,xt) \geq 1/n$. Hence, $x \in F_n^+$ and F_n^+ is closed. Similarly, F_n^- is closed yielding F_n closed.

Next, we demonstrate that each F_n is nowhere dense. Let $V \subset F_n^+$ be a nonempty open set and choose x from V. Since $x \in J^+(x)$ there are nets (x_i) and (x_it_i) in V with $x_i \to x$, $x_it_i \to x$, and $n \leq t_i \to +\infty$. However, $1/n \leq \rho(x_i,x_it_i) \to 0$ for each $\rho \in G$ which is impossible. Consequently, F_n^+ is nowhere dense. Similarly, F_n^- is nowhere dense for each n and we have each F_n nowhere dense.

Finally, the sets $X \setminus F_n$ are all dense open subsets of X. By the Baire property $\cap(X \setminus F_n) = X \setminus \cup F_n$ is dense in X. Moreover, the set $X \setminus \cup F_n$ consists of the Poisson stable points since a point x is not Poisson stable if and only if $x \in F_n$ for some n. The proof is complete.

As a result of this theorem we observe that whenever X is a completely regular (uniform) Baire space the set of Poisson stable points is dense in the set of central motions (in the sense of Birkhoff). The fact that the center M of a flow on a completely regular Baire space X contains the interior A^o of the set A of nonwandering points follows from $x \in J^+(x) \cap A^o = J_{A^o}^+(x)$ (the prolongation of x relatively to A^o) for each x in A^o (see 3.24.9 of [1]), a completely regular Baire subspace of X. Thus, $X = M \cup (\overline{X \setminus A})$ and we have the following proposition.

Proposition. If X is a completely regular Baire space, then the set consisting of the Poisson stable points and the wandering points is dense in X.

The set of central motions of a flow on a completely regular Baire space need not be the set of nonwandering points. Example 3.10 page 343 of [6] is such a flow. On the other hand, if X is non-Baire, the set of central motions need not be dense in X even though the flow is nonwandering as we see in the following example.

Example. The flow of Example 4.06 page 346 of [6] is defined on a torus T. The torus contains one critical point p. For exactly one orbit $C(x)$ and one orbit $C(y)$ we have $L^+(x) = \{p\} = L^-(y)$ and $L^-(x) = L^+(y) = T$. Every other orbit $C(z)$ is regular with $L^+(z) = L^-(z) = T$. Let $X = C(y)$. Then $X = C(y) = L^+_X(y)$ (the positive limit set of y relative to X) whereas $L^-_X(y) = \emptyset$. The space X is completely regular non-Baire and $(X, \pi | X)$ is a nonwandering flow with no Poisson stable points.

Nemytskii and Stepanov show that whenever X is compact metric the set of nonwandering points uniformly attract the wandering points. This is not generally true even for X locally compact. In [5] the statement is shown to be valid in the extended flow on the one point compactification, and hence, for X compact Hausdorff. Requiring that the nonwandering points positively and negatively attract the wandering points, i.e., no wandering point is unilaterally divergent, is not strong enough of a requirement to force the set of nonwandering points to uniformly attract the wandering points. Such flows on completely regular Baire spaces are easy to construct even when the phase space is not locally compact.

We conclude our remarks with a brief examination of orbits. The following are shown to be equivalent in [5] for $x \in X$: (a) $x \notin L(x)$; (b) π_x is a homeomorphism; and (c) $C(x)$ is homeomorphic to R. Using similar arguments the following can be shown to be equivalent for $x \in X$: (a) $x \notin L^+(x)$ $(x \notin L^-(x))$; (b) π_x^+ (π_x^-) is a homeomorphism, and (c) $C^+(x)$ $(C^-(x))$ is homeomorphic to $R^+(R^-)$. Also in [5] the condition $C(x)$ is a closed Baire space is shown to imply that $C(x)$ is homeomorphic to R, S^1, or a single point. Following parallel arguments we can show that $C^+(x)(C^-(x))$ a closed Baire space implies $C^+(x)(C^-(x))$ is homeomorphic to $R^+(R^-)$, S^1, or a single point.

The following theorem is given by Bhatia and Hajek in [1]. If X is locally compact or a complete metric space, then $\overline{L^+(x) \setminus C(x)} = L^+(x)$ $(\overline{L^-(x) \setminus C(x)} = L^-(x))$ for any nonperiodic noncritical point x in X, moreover, if x is positively (negatively) Poisson stable but nonperiodic noncritical, then $\overline{L^+(x) \setminus C(x)} = K(x)$ $(\overline{L^-(x) \setminus C(x)} = K(x))$. Their purpose for restricting spaces is to assure that $L^+(x)$ $(L^-(x))$ is a Baire subspace of X. Thus, these conclusions follow as long as $L^+(x)$ $(L^-(x))$ is a Baire subspace of X. Whenever $L^+(x)$ $(L^-(x))$ is not a Baire subspace of X we can have $\overline{L^+(x) \setminus C(x)}$ $(\overline{L^-(x) \setminus C(x)})$ empty or nonempty and unequal to $L^+(x)$ $(L^-(x))$ as well as have the conditions that hold when $L^+(x)$ $(L^-(x))$ is a Baire subspace of X. Examples of such flows are easily constructed from subspaces of the flow on the torus T given in Example 4.06 page 346 of [6].

REFERENCES

[1] Bhatia, N., and O. Hajek, *Theory of dynamical systems*, Part I, Tech. Note BN-599, Univ. of Maryland, 1969.

[2] Bhatia, N., and O. Hajek, *Local semi-dynamical systems*, Lect. Notes in Math. 90, Springer-Verlag, New York/Berlin 1969.

[3] Birkhoff, G., *Uber gewisse zentrale Bewegungen dynamischer systeme*, Kgl. Ges. d. Wiss. Gottingen, Nachrichten, Math. Phys. Kl., 1926.

[4] Birkhoff, G., "Dynamical Systems," Amer. Math. Soc. Coll. Publ., vol. IX, Prov., R.I., 1966.

[5] Knight, R., *Central motions* (to appear)

[6] Nemytskii, V., and V. Stepanov, "Qualitative Theory of Differential Equations," English Trans., Princeton, N.J. 1960.

ON MULTIPLE SOLUTIONS OF A NONLINEAR DIRICHLET PROBLEM

A. C. Lazer

University of Cincinnati

P. J. McKenna

University of Florida

INTRODUCTION

This paper is divided into three separate parts. Each part is related to a recent theorem of Berger [1] which concerns the nonlinear Dirichlet problem

$$Lu - \lambda u + g(x)u^{2p+1} = 0 \quad \text{in} \quad \Omega \subset \mathbb{R}^N \quad D^{\alpha}u|\partial\Omega = 0 \qquad (0.1)$$
$$|\alpha| \le m - 1 .$$

Berger showed that if $\lambda_1 < \lambda < \lambda_2$ where λ_1 and λ_2 are the first two eigenvalues of the elliptic operator L, if g is smooth and positive on the closure of the bounded region Ω, and that if p is a positive integer suitably restricted in terms of N and m, where $2m$ is the order of L, then (0.1) has exactly three solutions. It is implicitly assumed in [1] that L and $\partial\Omega$ are sufficiently regular.

In the first part we show that in the second-order case if the nonlinearity of (0.1) is replaced by a more general one, <u>which need not be odd</u>, and if $\lambda_2 < \lambda < \lambda_3$, then the problem has more than three solutions.

DIFFERENTIAL EQUATIONS

ISBN 0-12-045550-1

In the second part we use the methods of the first part to obtain a short proof of an extension of Berger's theorem. Whereas Berger's approach is to apply bifurcation theory and results on proper mappings to an abstract equation, our approach is to calculate the Leray-Schauder indices of the zeros of a certain vector field and then to use a global theorem concerning the sum of these indices.

In the third part we briefly show how an abstract result due to Clark [2] can be used to give a lower bound on the number of solutions of class of boundary value problems containing (0.1). This bound is expressed in terms of the number of eigenvalues of L which are strictly less than λ.

1. We begin by considering the second-order Dirichlet problem

$$\Delta u + \lambda u - g(x,u)u = 0 \quad \text{in } \Omega \quad u \mid \partial\Omega = 0. \tag{1}$$

We make the following assumptions on λ, g and Ω:

A (i): $\lambda_2 < \lambda < \lambda_3$ where $0 < \lambda_1 < \lambda_2 \leq \lambda_3 \leq \ldots$ are the eigenvalues of $-\Delta$ with the Dirichlet boundary condition. In particular, this requires λ_2 to be simple.

A (ii): Both g and $\frac{\partial g}{\partial u}$ are continuous on $\overline{\Omega} \times (-\infty, \infty)$ and satisfy a Hölder condition with exponent $\alpha \in (0, 1)$ in x and uniform Hölder constants for u in bounded intervals. The region Ω is bounded in $\mathbb{R}^N$ $(N \geq 1)$ and $\partial\Omega$ is of class $C^{2+\alpha}$.

A (iii): $g(x, 0) = 0$ for all $x \in \Omega$, $\frac{\partial g}{\partial u}(x, u)\, u > 0$ for $u \neq 0$ and $x \in \overline{\Omega}$, and $\lim_{u \to \infty} g(x, u) > \lambda$, $\lim_{u \to -\infty} g(x, u) > \lambda$ uniformly with respect to $x \in \overline{\Omega}$, where either limit may be $+\infty$.

We emphasize that g need not be even in u.

The following theorem will be valid if Δ is replaced by any second-order, uniformly elliptic self adjoint operator for which the maximum principle is valid; we have chosen Δ notational convenience in what follows.

Theorem 1. *Under assumptions A (i) - A (iii) the boundary value problem (1) has at least four solutions - at least five solutions if $g(x, u)$ is assumed to be an even function of u. Moreover, (1) has a solution which is strictly positive in Ω and a solution which is strictly negative in Ω.*

That (1) has five solutions in the odd case also essentially follows from results in [4].

To prove Theorem 1, we begin with a truncation argument in order to make the problem (1) amenable to the methods of Leray-Schauder degree theory. By A (iii) there exist numbers b and c with $b < 0 < c$ such that $g(x, u) > \lambda$ if $x \in \overline{\Omega}$ and either $u \leq b$ or $u \geq c$. Let φ_k, $k = 1, 2$, be the real-valued functions defined on $(-\infty, \infty)$ by
$\varphi_1(t) = b + \arctan(u - b)$ and $\varphi_2(t) = c + \arctan(u - c)$.
We define $h : \overline{\Omega} \times (-\infty, \infty) \to \mathbb{R}$ by $h(x, u) = g(x, \varphi_1(u))$ if $x \in \overline{\Omega}$ and $u < b$, $h(x, u) = g(x, u)$ if $x \in \overline{\Omega}$ and $b \leq u \leq c$ and $h(x, u) = g(x, \varphi_2(u))$ if $x \in \Omega$ and $c < u$. It is easy to check the continuity of h and $\frac{\partial h}{\partial u}$. Moreover, we have

$$h(x, 0) = 0, \qquad \frac{\partial h}{\partial u}(x, u)u > 0, \ u \neq 0 \tag{2}$$

and

$$h(x, u) > \lambda \quad \text{if} \quad u \geq c \quad \text{or} \quad u \leq b. \tag{3}$$

We consider the boundary value problem

$$\Delta u + \lambda - h(x, u)u = 0 \quad \text{in } \Omega \qquad u \,|\, \partial\Omega = 0. \tag{4}$$

Since (3) implies that $h(x, u) - \lambda u > 0$ if $u \geq c > 0$, and $h(x, u)u - \lambda u < 0$ if $u \leq b < 0$ it follows from the maximum principle that if $u(x)$ is any solution of (4) then $b \leq u(x) \leq c$, for all $x \in \Omega$. Therefore, from the definition of h we have

Lemma 1. *Any classical solution of the boundary value problem (4) is also a solution of the boundary value problem (1).*

Since h is bounded on $\bar{\Omega} \times (-\infty, \infty)$, it follows from standard results (see, for example [6]) that $v(x) \to h(x, v(x))\, v(x)$ defines a continuous mapping from $L^2(\Omega)$ into itself which takes bounded sets into bounded sets. As is well known, Δ can be extended to a topological isomorphism from the Sobolev space $\overset{0}{H}_{2,2}(\Omega)$ onto $L^2(\Omega)$. Therefore, by the continuity and compactness of the injections $\overset{0}{H}_{1,2}(\Omega) \to L^2(\Omega)$ and $\overset{0}{H}_{2,2}(\Omega) \to \overset{0}{H}_{1,2}(\Omega)$ the mapping $F : \overset{0}{H}_{1,2}(\Omega) \to \overset{0}{H}_{1,2}(\Omega)$ defined by

$$F(u) = \Delta^{-1}[h(x, u)u - \lambda u] \qquad (5)$$

is completely continuous.

Lemma 2. *There exists a number* $r > 0$, *independent of* $s \in [0, 1]$, *such that if* $u \in \overset{0}{H}_{1,2}(\Omega)$ *and* $u = s\,F(u)$ *then* $\|u\|_1 < r$ *where* $\| \ \|_1$ *denotes the usual norm in* $\overset{0}{H}_{1,2}(\Omega)$. *If* B *denotes the set of* $u \in \overset{0}{H}_{1,2}(\Omega)$ *satisfying* $\|u\| < r$ *then* $d(u - F(u), B, 0) = 1$.

Proof: According to (3), there exist a number $c_1 > 0$ such that $h(x, u)u^2 - \lambda u^2 \geq -c_1$ for all $(x, u) \in \Omega \times (-\infty, \infty)$. If $s \in (0, 1]$ and $u = s\,F(u)$ then u is a weak $\overset{0}{H}_{1,2}(\Omega)$-solution of the boundary value problem

$$\Delta u + s\,[\lambda u - h(x,\ u)u] = 0 \qquad u \mid \Omega = 0. \quad \text{Therefore,}$$

$$\int_{\Omega} (|\text{grad } u|^2 + s(h(x,\ u) - \lambda)u^2))dx = 0$$

so

$$\int_{\Omega} |\text{grad } u|^2\, dx \le c_1 \ \text{meas}\ \Omega\ .$$

Hence,

$$\|u\|_1^2 = \int_{\Omega}(|\text{grad } u|^2 + u^2)dx \le (1 + \lambda_1^{-1})c_1 \ \text{meas}\ \Omega$$

and the first part of the lemma follows by setting

$r^2 = 1 + (1 + \lambda_1^{-1})c_1$ meas Ω .

To prove the second part we note that, by the invariance under homotopy property of Leray-Schauder degree, for $s \in [0,\ 1]$, $d(u - s\,F(u),\ B,\ 0)$ = constant = $d(u,\ B,\ 0) = 1$.

<u>Lemma 3</u>. <u>There exist solutions</u> u_1 <u>and</u> u_2 <u>of</u> (4) <u>such that</u> $u_1(x) < 0 < u_2(x)$ <u>for all</u> $x \in \Omega$. <u>Moreover,</u> u_1 <u>and</u> u_2 <u>are nonsingular solutions of</u> $u - F(u) = 0$ <u>with</u> $i(u - F(u),\ u_1,\ 0) = i(u - F(u),\ u_2,\ 0) = 1$.

<u>Proof</u>: To establish the existence of u_2 we consider the boundary value problem

$$\Delta u + \lambda u - h_2(x,\ u)u = 0,\ u|\partial\Omega = 0 \tag{6}$$

where h_2 is defined by

$$h_2(x,\ u) = h(x,\ |u|)\ . \tag{7}$$

It is easy to check that the product $h_2(x,\ u)u$ has a continuous derivative with respect to u for $(x,\ u) \in \overline{\Omega} \times (-\infty,\infty)$ and that (3) holds with h replaced by h_2 . If

$$G_2(x,\ u) = \int_0^u h_2(x,\ t)\ t\ dt$$

then $G_2(x,\ u) = G_2(x,\ -u)$ and (3) implies the existence of a constant c_2 such that $G_2(x,\ u) - \frac{\lambda u^2}{2} \ge c_2)$ for all

$(x, u) \in \overline{\Omega} \times (-\infty, \infty)$. Therefore, if we define $J_2 : \overset{0}{H}_{1,2}(\Omega) \to \mathbb{R}$ by

$$J_2[u] = \int_\Omega \left(\frac{|\text{grad } u|^2}{2} + G_2(x, u) - \frac{\lambda u^2}{2}\right) dx$$

it follows that $J_2[u] \to \infty$ as $\|u\|_1 \to \infty$. The fact that J_2 is defined follows from the boundedness of h_2. Since J_2 can be expressed as the sum of a convex function and a function that is continuous with respect to weak convergence, it follows that J_2 is lower semi-continuous with respect to weak convergence. Therefore, there exists a $u_0 \in \overset{0}{H}_{1,2}(\Omega)$ such that $J_2[u_0] = \min J_2[u]$. Since $\lambda > \lambda_2$, $u_0 \neq 0$. Indeed, since $h_2(x,0) = 0$,

$G_2(x,0) = \frac{\partial G_2}{\partial u}(x, 0) = \frac{\partial^2 G_2}{\partial u^2}(x, 0) = 0$; therefore, if φ_1 is a normalized eigenfunction of $-\Delta$ corresponding to the eigenvalue λ_1 and $\varepsilon > 0$ is sufficiently small then

$$J_2[\varepsilon\varphi] = \int_\Omega \left(\varepsilon^2 \frac{\lambda_1}{2} \varphi_1^2 - \frac{\lambda \varphi_1^2 \varepsilon^2}{2} + G(x, \varepsilon \varphi_1)\right) dx$$

$$\leq \frac{1}{4} \varepsilon^2 (\lambda_1 - \lambda) < 0 = J_2[0] .$$

If $u_2(x) = |u_0(x)|$ then standard methods - for example, the use of difference-quotients - show that $u_2 \in \overset{0}{H}_{1,2}(\Omega)$ and that

$$\int_\Omega |\text{grad } u_2|^2 \, dx \leq \int_\Omega |\text{grad } u_0|^2 \, dx .$$

Thus, since $G_2(x, u)$ is an even function of u, $J_2[u_2] = J_2[u_0] = \min J_2[u]$. It follows that for any $w \in \overset{0}{H}_{1,2}(\Omega)$, we have $0 = J_2'[u_2](w) =$

$$\int_\Omega ((\text{grad } u_2 , \text{grad } w) - \lambda u_2 w + h_2(x, u_2)u_2 w) \, dx .$$

This means that u_2 is a weak solution of (6) and by standard regularity theory u_2 is a classical solution of (6). Since $u_2(x) \geq 0$ for all $x \in \Omega$ and $h_2(x, u) = h(x, u)$ for $u \geq 0$, it follows that u_2 is a solution of (4). We assert that $u_2(x) > 0$ for $x \in \Omega$. To see this we note that if $v = -u_2$ then $\Delta v - h(x, u_2)v = \lambda u_2 \geq 0$ in Ω. Since $h(x, u_2) \geq 0$, $v \mid \partial\Omega = 0$, and $v \not\equiv 0$, the maximum principle implies that $v(x) < 0$ for all $x \in \Omega$. This proves the assertion.

To calculate $i(u - F(u) , u_2 , 0)$ we first note that since u_2 is a solution of the linear problem $\Delta w + \lambda w - h(x, u_2)w = 0$, $w \mid \partial \Omega = 0$ and since $u_2(x) \neq 0$ for all $x \in \Omega$, $\alpha = 0$ must be the lowest eigenvalue of the eigenvalue problem

$$\Delta w + \lambda w - h(x, u_2)w + \alpha w = 0 , \quad w \mid \partial \Omega = 0 . \tag{8}$$

Therefore, according to the well-known variational characterization of the lowest eigenvalue of (8), it follows that

$$\int_\Omega (|\text{grad } v|^2 + h(x, u_2)v^2 - \lambda v^2) \, dx \geq 0 \tag{9}$$

for all $v \in \overset{0}{H}_{1,2}(\Omega)$. To show that u_2 is a nonsingular solution of $u - F(u) = 0$ and that $i(u - F(u), u_2, 0) = 1$, it is sufficient to show that there exists no $v \in \overset{0}{H}_{1,2}(\Omega)$ such that $v \neq 0$ and $\gamma v = F'(u_2)v$ for some $\gamma \geq 1$ - see, for example [5, pp. 66-67]. Assuming the contrary, it would follow from (5) that for some $v \neq 0$ and some $\gamma \geq 1$ $\gamma v = \Delta^{-1}[h(x, u_2)v + \frac{\partial h}{\partial u}(x, u_2)u_2 v - \lambda v]$ and hence that

$$\gamma\Delta v + \lambda v - (h(x, u_2) + \frac{\partial h}{\partial u}(x, u_2)u_2)v = 0, \quad v \mid \partial\Omega = 0.$$

But, since $\gamma \geq 1$, it would follow from (2) that

$$\int_\Omega (|\text{grad } v|^2 + h(x, u_2)v^2 - \lambda v^2)\, dx <$$

$$\int_\Omega (\gamma \mid \text{grad } v|^2 + (h(x, u_2) + \frac{\partial h}{\partial u}(x, u_2)u_2)v^2 - \lambda v^2)dx = 0.$$

Since this contradicts (9), the proof that $i(u - F(u), u_2, 0) = 1$ is complete.

The existence of a solution u_1 of (4) with $u_1(x) < 0$ for all $x \in \Omega$ can be established by considering the functional

$$J_1[u] = \int_\Omega (\frac{|\text{grad } u|^2}{2} - G_1(x, u) - \frac{\lambda}{2}u^2)dx, \quad u \in \overset{0}{H}_{1,2}(\Omega),$$

where

$$G_1(x, u) = \int_0^u h(x, -|t|)t\, dt.$$

If $\hat{u}_0$ is an element in $\overset{0}{H}_{1,2}(\Omega)$ which minimizes J_1, then since $G_1(x, u)$ is an even function of u, it follows that the function $u_1 = -|\hat{u}_0|$ also minimizes J_1. Moreover, the same argument which was applied to the functional J_2 shows that $u_1 \neq 0$ and that u_1 is a solution of (4). The proof that u_1 is strictly negative and that $i(u - F(u), u_1, 0) = 1$ follows from the same reasoning that was used above.

To complete the proof of Theorem 1 we note that $u = 0$ is a solution of (3). To calculate the index of $u - F(u)$ at 0 we observe that since $h(x, 0) = 0$, it follows from (5) that for any $v \in \overset{0}{H}_{1,2}(\Omega)$, $F'(0)v = -\Delta^{-1}\lambda v = -\lambda\Delta^{-1}v$. It follows that if $v \neq 0$ and $F'(0)v = \gamma v$ then $\gamma\Delta v + \lambda v = 0 \quad v \mid \partial\Omega = 0$ so $\gamma = \lambda / \lambda_k$ for some $k \geq 1$. Since $\lambda_2 < \lambda < \lambda_3$ we see that 0 is a nonsingular solution of $u - F(u) = 0$ and that $F'(0)$ has exactly two eigenvalues greater than one. Hence, $i(u - F(u), 0, 0) = (-1)^2 = 1$.

In summary we have established the existence of three nonsingular solutions of $u - F(u) = 0$, each with index 1. If u_1, u_2 and 0 were the only solutions of $u - F(u) = 0$ then for B as in Lemma 2, it would follow from degree theory that

$$d(u - F(u), B, 0) = i(u - F(u), 0, 0) + i(u - F(u), u_1, 0) + i(u - F(u), u_2, 0) = 3$$

which contradicts Lemma 1. This contradiction shows that $u - F(u) = 0$, and hence (4) has at least four solutions. The first part of Theorem 1 now follows from Lemma 1.

If $g(x, u)$ is an even function of u then if v is a solution of (1), so is $-v$. Since (1) has at least three nonzero solutions, it follows that if g is even in u, then (1) must have at least four nonzero solutions and therefore at least five solutions.

Remark: By examining the proof of Theorem 1, it is easy to see that the same result holds if the condition $\lambda_2 < \lambda < \lambda_3$ is replaced by $\lambda_{2k} < \lambda < \lambda_{2k+1}$ for some $k > 1$.

2. We now consider the boundary value problem

$$Lu - \lambda u + g(x,u)u = 0 \text{ in } \Omega \quad D^\alpha u \mid \partial\Omega = 0 \qquad (10)$$
$$|\alpha| \le m - 1 .$$

Here again Ω is a bounded open subset of $\mathbb{R}^N$ and L is a selfadjoint, strongly elliptic operator of order $2m$ with smooth coefficients and with eigenvalues

$$\lambda_1 \le \lambda_2 \le \dots \le \lambda_k \to \infty .$$

Here we shall only consider the question of weak

$\overset{0}{H}_{m,2}(\Omega)$-solutions. With enough smoothness assumptions on L, g, and $\partial\Omega$, regularity of such solutions may be established by standard bootstrap arguments.

Theorem 2. *Let* g *and* $\frac{\partial g}{\partial u}$ *be continuous on* $\bar{\Omega} \times (-\infty, \infty)$ *and assume that* $g(x, 0) = 0$ *and* $u \frac{\partial g}{\partial u}(x, u) > 0$ *for* $u \neq 0$. *Assume that* g *and* $\frac{\partial g}{\partial u}$ *satisfy growth conditions of the form*

$$|g(x, u)| \le a_1 + a_2 |u|^s \tag{11}$$

$$\left|\frac{\partial g}{\partial u}(x, u)\right| \le b_1 + b_2 |u|^{s-1}$$

where s *is arbitrary if* $2m \ge N$ *or* $s < 4m/(N - 2m)$ *if* $2m < N$. *If* $\lambda_1 < \lambda < \lambda_2$ *and* $\lim_{u \to -\infty} g(x, u) > \lambda$, $\lim_{u \to +\infty} g(x, u) > \lambda$ *uniformly with respect to* $x \in \bar{\Omega}$, *then* (10) *has exactly three solutions*. (As before, the limits may be infinite.)

To prove this theorem, we first observe, as in [1], that if Q is the symmetric bilinear form defined on $\overset{0}{C}{}^{\infty}(\Omega) \times \overset{0}{C}{}^{\infty}(\Omega)$ by $Q(u, v) = \int_{\Omega} u\, L\, v\, dx$ then it may be assumed by replacing L and λ by $L + C$ and $\lambda + C$ respectively, for suitable C, that $k|v|_m^2 \le Q(v, v)$ where $k > 0$ and $|\ |_m$ denotes the usual norm on $\overset{0}{H}_{m,2}(\Omega)$. Therefore, if Q is extended by continuity to $\overset{0}{H}_{m,2}(\Omega) \times \overset{0}{H}_{m,2}(\Omega)$, the same inequality will be valid for $v \in \overset{0}{H}_{m,2}(\Omega)$.

We consider the case $2m < N$. Since $s < 4m/(N - 2m)$ we see that $2N(s + 1) / (N + 2m) < 2N / (N - 2m)$ and therefore, by the Sobolev imbedding theorem, *the inclusion map from*

$\overset{0}{H}_{m,2}(\Omega)$ into $L_{\frac{2N(s+1)}{N+2m}}(\Omega)$ <u>is compact</u>. The condition (11) implies the existence of constants d_1 and d_2 such that

$$|\lambda u - g(x, u)| \le d_1 + d_2 |u|^{s+1}$$

so, by a theorem of Vainberg (see [5]), $u \to g(x, u)u - \lambda u$ <u>defines a continuous mapping from</u> $L_{\frac{2N(s+1)}{N+2m}}(\Omega)$ <u>into</u> $L_{\frac{2N}{N+2m}}(\Omega)$ <u>which takes bounded sets into bounded sets</u>. Next, we observe that, since L has no zero eigenvalues, standard elliptic theory shows that L defines a continuous, linear, one-to-one mapping of the Sobolev space $\overset{0}{H}_{2m,\frac{2N}{N+2m}}(\Omega)$ onto $L_{\frac{2N}{N+2m}}(\Omega)$; in (12) below, L^{-1} will denote the inverse of this mapping. As Sololev's imbedding theorem shows that the inclusion of $\overset{0}{H}_{2m,\frac{2N}{N+2m}}(\Omega)$ in $\overset{0}{H}_{m,2}(\Omega)$ is continuous, the above considerations show that if

$$G(u) = L^{-1}[\lambda u - g(x, u)\, u] \tag{12}$$

then, <u>regarded as a mapping from</u> $\overset{0}{H}_{m,2}(\Omega)$ <u>into itself</u>, G <u>is completely continuous</u>.

If $2m \ge N$ then for any $q \ge 1$, the inclusion of $\overset{0}{H}_{m,2}(\Omega)$ in $L_q(\Omega)$ is compact. Therefore, the above argument applied to the factorization

$$\overset{0}{H}_{m,2}(\Omega) \subset L_{2(s+1)}(\Omega) \xrightarrow{\lambda u - gu} L_2(\Omega) \xrightarrow{L^{-1}} \overset{0}{H}_{2m,2}(\Omega) \subset \overset{0}{H}_{m,2}(\Omega)$$

again shows that G is completely continuous.

The conditions of Theorem 2 imply the existence of a number $d \geq 0$ such that $g(x, u)u^2 - \lambda u^2 \geq -d$ for all $(x, u) \in \overline{\Omega} \times (-\infty, \infty)$. From this it follows that if $s \in (0, 1]$ and if $u \in \overset{0}{H}_{m,2}(\Omega)$ satisfies $u = s\,G(u)$ then

$$k|u|_m^2 - d \text{ meas } \Omega \leq Q(u, u) + s \int_\Omega (g(x, u)u^2 - \lambda u^2)\, dx = 0$$

so the same argument used to prove Lemma 2 yields

Lemma 4. *There exists a number* $r > 0$ *such that* $u = G(u)$ *implies* $|u|_m < r$. *If* $D = \{u \in \overset{0}{H}_{m,2}(\Omega) \mid |u|_m < r\}$ *then* $d(u - g(u), D, 0) = 1$.

In [1] an eigenvalue comparison argument is used to show that every nonzero solution of the problem considered there is, in a certain sense, nonsingular. Here we use a similar argument to establish a stronger result.

Lemma 5. If $u_0 \neq 0$ *and* $u_0 = G(u_0)$ *then* $i(u - G(u), u_0, 0) = 1$.

Proof: According to (12), it follows that if $v \in \overset{0}{H}_{m,2}(\Omega)$ then

$$G'(u_0)(v) = L^{-1}\left[\lambda v - g(x, u_0)v - \frac{\partial g}{\partial u}(x, u_0)u_0 v\right]. \quad (13)$$

The Frechet differentiability of G and calculation of $G'(u_0)$ follows from a routine but lengthy application of Vainberg's theorem, Holder's inequality and (11) (See [4 p. 987]) where a similar calculation is given. Since u_0 is a solution - in the weak sense - of the linear problem

$$Lv - \lambda v + g(x, u_0)v - \alpha v = 0 \qquad D^\sigma v \mid \partial\Omega = 0 \quad (14)$$

$$|\sigma| \leq m - 1$$

when $\alpha = 0$, it follows that 0 is an eigenvalue of the problem (14). Since $g(x, u_0) \geq 0$, it follows from the well-known variational characterization of eigenvalues in terms of Rayleigh quotients that if $\{\alpha_n\}_1^\infty$ denotes the sequence of ordered eigenvalues of problem (14) and if $\{\beta_n\}_1^\infty$ denotes the sequence of ordered eigenvalues of the problem

$$L v - \lambda v - \beta v = 0 \tag{15}$$

$$D^\alpha v \mid \partial\Omega = 0 \qquad |\alpha| \leq m - 1$$

then $\beta_n \leq \alpha_n$ for all n. Since $\beta_2 = \lambda_2 - \lambda > 0$, we see that $\alpha_1 = 0$. Thus, it follows from the variational characterization of the lowest eigenvalue of (14) that for arbitrary $w \in \overset{0}{H}_{m,2}(\Omega)$

$$Q(w, w) + \int_\Omega (g(x, u_0)w^2 - \lambda w^2)\, dx \geq 0 . \tag{16}$$

As with Lemma 3, Lemma 5 will follow from the fact that there exists no $w_0 \in \overset{0}{H}_{m,2}(\Omega)$ such that $w_0 \neq 0$ and $\gamma w_0 = G'(u_0)w_0$ for some $\gamma \geq 1$. Assuming the contrary, it follows from (13) that

$$\gamma L w_0 - \lambda w_0 + g(x, u_0)w_0 + \frac{\partial g}{\partial u}(x, u_0)u_0\, w_0 = 0$$

$$D^\alpha w_0 \mid \partial\Omega = 0 \qquad |\alpha| \leq m - 1$$

in the weak sense. Therefore,

$$Q(w_0, w_0) + \int_\Omega (g(x, u_0) - \lambda)w_0^2\, dx \leq$$

$$\gamma Q(w_0, w_0) + \int_\Omega (g(x, u_0) + \frac{\partial g}{\partial u}(x, u_0)u_0 - \lambda)\, w_0^2\, dx = 0 .$$

From (16) it follows that $\gamma = 1$ and that $\frac{\partial g}{\partial u}(x, u_0)u_0\, w_0^2 = 0$ a. e. on Ω which, according to the hypothesis of Theorem 2, implies that $g(x, u_0)\, w_0 = 0$ a. e.

on Ω . According to (16), w_0 minimizes the functional

$$Q(w, w) + \int_\Omega (g(x, u_0)w^2 - \lambda w^2)\, dx$$

on $\overset{0}{H}_{m,2}(\Omega)$ so by considering the Euler-Lagrange equation we see that

$$0 = L w_0 - \lambda w_0 + g(x, u_0)w_0 = L w_0 - \lambda w_0 \quad \text{in } \Omega$$

$$D^\alpha w_0 \mid \partial\Omega = 0 \qquad |\alpha| \le m - 1$$

(in the weak sense). Since this contradicts the condition $\lambda_1 < \lambda < \lambda_2$ the lemma is proved.

The proof of Theorem 2 is now concluded in the same manner as the proof of Theorem 1. The hypotheses of Theorem 1 imply that $G(0) = 0$ and $G'(0)v = \lambda L^{-1}v$. Hence, the eigenvalues of $G'(0)$ are the numbers λ / λ_k , $k \ge 1$, and since $\lambda / \lambda_1 > 1 > \lambda / \lambda_2$ we see that $i(u - G(u), 0, 0) = -1$. Therefore, since every nonzero solution has index equal to $+1$, it follows from Lemma 4 and a fundamental result of degree theory that $u - G(u) = 0$ has only a finite number of solutions and that

$$1 = d(u - g(u), D, 0) = -1 + \text{number of nonzero solutions.}$$

The assertion of Theorem 2 follows from this equation.

3. Finally, we consider briefly the problem (10) with g an even function of u and $\lambda_k < \lambda \le \lambda_{k+1}$ where $k \ge 1$ is arbitrary. In this case the conditions on g can be weakened considerably.

<u>Theorem 3</u>. <u>Let</u> g <u>be continuous on</u> $\bar{\Omega} \times (-\infty, \infty)$ <u>with</u> $g(x, u) \equiv g(x, -u)$ <u>and</u> $g(x, 0) = 0$. <u>Let</u> $\lambda_k < \lambda \le \lambda_{k+1}$ <u>where</u> $k \ge 1$. <u>If</u> g <u>satisfies the first growth condition in</u> (11) <u>where</u> s <u>is as in Theorem</u> 2 <u>and</u> $\lim_{u \to \infty} g(x, u) > \lambda$

<u>uniformly in</u> x <u>then</u> (1) <u>has at</u> k <u>distinct pairs</u> $\pm u$ <u>of nontrivial weak</u> $\overset{0}{H}_{m,2}(\Omega)$ - <u>solutions</u>. As in the case of Theorem 1, regularity can be established with more smoothness assumptions.

In the second order case the theorem was established by Hempel [4] but with more assumptions than were necessary on g - for example, $g(x, u)u > 0$, $u \neq 0$, and $g(x, u)$ increasing in u on $[0, \infty)$. Theorem 3 may be obtained by considering the functional

$$f(u) = Q(u, u) + \int_\Omega (2\, F(x, u(x)) - \lambda u(x)^2)\, dx$$

where

$$F(x, u) = \int_0^u g(x, t)\, t\, dt.$$

That f is defined on $\overset{0}{H}_{m,2}(\Omega)$ in case $2m < N$ follows from the estimate $|F(x, u)| \leq k_1 + k_2 |u|^{s+2}$ the inequality $s + 2 < 2N/(N - 2m)$, and the compactness of the inclusion of $\overset{0}{H}_{m,2}(\Omega)$ in $L_{s+2}(\Omega)$. We note that $f(-u) = f(u)$, $f(u) \to \infty$ as $|u|_m \to \infty$, and that $(f(u) - B(u))/|u|_m^2 \to 0$ as $|u|_m \to 0$ where

$$B(u) = Q(u, u) - \int_\Omega \lambda\, u(x)^2\, dx .$$

The hypotheses of the theorem imply that the quadradic form B is negative definite on the k-dimensional subspace of $\overset{0}{H}_{m,2}(\Omega)$ spanned by the eigenfunctions of L corresponding to $\lambda_1, \ldots, \lambda_k$. Since it is not difficult to show that f is Frechet - differentiable and satisfies a certain technical condition known as the Palais-Smale condition, Theorem 3 follows from an abstract result due to D. Clark. (See [2, Theorem 11].) We refer the reader to [3], where Clark has applied this abstract result to a similar problem concerning periodic solutions of Hamiltonian systems of differential equations, for more details.

REFERENCES

[1] Berger, M. S., *Nonlinear problems with exactly three solutions*, Indiana Univ. Math. J., 28(1979), pp. 689-698.

[2] Clark, D. C., *A variant of Lusternik-Schnivel man theory*, Indiana Univ. Math. J., 22(1972), pp. 65-74.

[3] Clark, D. C., *On periodic solutions of autonomous Hamiltonian systems of ordinary differential equations*, Proc. Am. Math. Soc., 39(1973), pp. 579-584.

[4] Hempel, J. A. *Multiple solutions for a class of nonlinear boundary value problems*, Indiana Univ. Math. J. 20(1971), pp. 983-996.

[5] Nirenberg, L., Topics in Nonlinear Functional Analysis, New York kniversity, 1974.

[6] Vainberg, M. M., Variational Methods for the Study of Nonlinear Operators, Holden-Day, San Francisco 1964.

CERTAIN "NONLINEAR" DYNAMICAL SYSTEMS ARE LINEAR

Roger C. McCann

Mississippi State University

INTRODUCTION

Suppose that we have dynamical system π on R^n which has a globally asymptotically stable critical point p. It is well known that there is a Liapunov function L for π. If $\lambda > 0$ is sufficiently small, then $L^{-1}(\lambda)$ is a compact section for π restricted to $R^n - \{p\}$, i.e., there is a continuous mapping $\tau : R^n - \{p\} \to R$ such that $\pi(x,\tau(x)) \varepsilon L^{-1}(\lambda)$. Since $S \equiv L^{-1}(\lambda) \subset R^n$ and the one point compactifaction of R^n is homeomorphic to S^n, the unit sphere in R^{n+1}, there is an embedding $g : S \to S^n \subset R^{n+1}$.

Consider the dynamical system ρ defined on R^{n+1} by the linear differential equation

$$\frac{dy}{dt} = -y.$$

Then $\rho(x,t) = e^{-t}x$ for every $(x,t) \varepsilon R^n \times R$. Define $G : R^{n+1}$ by

$$G(x) = \begin{cases} 0 & \text{if } x = p \\ e^{-\tau(x)}g(\pi(x,\tau(x))) & \text{if } x \varepsilon R^{n+1} - \{p\}. \end{cases}$$

ISBN 0-12-045550-1

It can be shown that G is a homeomorphism of R^n onto $G(R^n)$ and $G(\pi(x,t)) = \rho(G(X),t)$. Thus, the dynamical system π can be embedded into a dynamical system ρ on R^{n+1} which is linear in the sense that $\rho(\cdot,t)$ is a linear mapping of R^{n+1} onto R for every $t \in R$.

This idea can be generalized to dynamical systems on separable metric spaces. However, in such a case we need a stronger stability concept than asymptotic stability.

Definition. A compact subset M of a topological space X is uniformly asymptotically stable with respect to a dynamical system π on X if there is a neighborhood U of M such that for any neighborhood $V \subset U$ of M there is a $T > 0$ such that $\pi(U,[T,\infty)) \subset V$.

It is not difficult to show that if X is locally compact then the concepts of asymptotic stability and uniform asymptotic stability coincide.

If π is a dynamical system on a separable metric space X and there is a globally uniformly asymptotically stable critical point, then we shall embed π into the dynamical system ρ on ℓ_2 defined by $\rho(x,t) = e^{-t}x$. This can be done as follows. Set $S = L^{-1}(\lambda)$ where L is a Liapunov function for π and λ is a number in the range of L such that $L^{-1}(\lambda) \subset U$($U$ a neighborhood of p as in the definition of uniform asymptotic stability). Then S is a section for π restricted to $X - \{p\}$ and there is a continuous mapping $\tau : X - \{p\} \to R$ such that $\pi(x,\tau(x)) \in S$. Let $\{x_n\}$ be a countable dense subset of S and define a countable number of continuous mappings $f_n : S \to R$ by $f_n(x) = d(x,x_n)$ where

d is a metric on X such that $d(x,y) \leq 1$ for every $x,y \in X$. It can be shown that the mapping $h : S \to \ell_2$ defined by

$$h(x) = (f_1(x), \tfrac{1}{2} f_2(x), \ldots, \tfrac{1}{n} f_n(x), \ldots)$$

is a homeomorphism of S onto $h(S)$ and that the mapping $H : X \to \ell_2$ defined by

$$H(x) = \begin{cases} 0 & \text{if } x = p \\ e^{-\pi(x)} h(\pi(x,\tau(x))) & \text{if } x \in X - \{p\} \end{cases}$$

is a homeomorphism of X onto $H(X)$ such that $H((x,t)) = e^{-t}H(x)$. Thus, we have embedded the dynamical system π into the dynamical system ρ on ℓ_2. This outlines the proof of the following theorem.

Theorem 1. Let π be a dynamical system on a separable metric space X which has a globally asymptotically stable critical point p and let ρ be the dynamical system defined on ℓ_2 by $\rho(x,t) = e^{-t}x$. Then π can be embedded into ρ if and only if p is uniformly asymptotically stable.

It is also possible to embed certain other dynamical systems, which have globally uniformly asymptotically stable compact sets, into a dynamical system $\hat{\pi}$ on ℓ_2 which is linear in the sense that $\hat{\pi}(\cdot,t)$ is a linear mapping for every $t \in R$. The following theorem was done in collaboration with L. Janos and J. Solomon.

Theorem 2. Let π be a dynamical system on a separable metric space X satisfying the conditions

(i) The mapping $\pi(\cdot,t) : X \to X$ is nonexpansive for every $t \geq 0$, i.e., if d is the metric on X, then $d(\pi(x,t),\pi(y,t)) \leq d(x,y)$ for every $x,y \in X$ and $t \geq 0$,

(ii) There is a compact invariant subset M of X which is globally uniformly asymptotically stable.

Then π can be embedded into a dynamical system $\hat{\pi}$ on ℓ_2 which is linear in the sense that $\hat{\pi}(\cdot,t)$ is a linear mapping for every $t \in R$.

Example. Let X be the plane R^2 with the open unit disc removed and let the dynamical system π on X be determined by the following autonomous system of polar differential equations

$$\frac{dr}{dt} = r(1 - r)$$

$$\frac{d\theta}{dt} = 1$$

The unit circle C is a periodic trajectory with all other trajectories spiraling toward it as $t \to \infty$. Hence C is uniformly asymptotically stable. It is also evident that the action of π is nonexpansive relative to the Euclidean metric on X. Therefore, by Theorem 2, π can be embedded into a linear dynamical system on ℓ_2.

The gist of the proof of the Theorem 2 is as follows. Identifying the set M to a point we form the quotient space X/M which is again separable, metrizable, and carries a dynamical system π^* induced by π. Moreover π^* satisfies the hypotheses of Theorem 1. It can be shown that there is a retraction $a : X \to M$ which commutes with $\pi(\cdot,t)$ for every $t \in R$. With α the natural projection on X onto X/M, an embedding i of X into $X/M \times M$ can be defined $i(x) = (\alpha(x), a(x))$. Since $\pi(\cdot, t)$ commutes with both α and a for every $t \in R$, π determines a dynamical system on $X/M \times M$ defined by $\pi^* \times \pi|M$ into which the mapping i embeds π. The first factor can be embedded into a linear dynamical system on ℓ_2 by Theorem 1. It can also be shown, using theorems concerning the linearization of locally compact transformation groups, that the second factor can also be embedded into a linear dynamical system on ℓ_2. Since $\ell_2 \times \ell_2$ can be identified with ℓ_2, the dynamical system $\hat{\pi}$ can be embedded into a linear dynamical system on ℓ_2.

REFERENCES

[1] Janos, L., R. McCann, J. Solomon, *Nonexpansive uniformly asymptotically stable flows are linear* (submitted for publication)

[2] McCann, R., *Embedding asymptotically stable dynamical systems into radial flows in* ℓ_2, (to appear in Pacific J. Math.)

[3] McCann, R., *Asymptotically stable dynamical systems are linear,* Pacific J. Math., 81(1979), pp. 475-479.

A MODEL OF COMPLEMENT ACTIVATION BY ANTIGEN-ANTIBODY COMPLEXES

Stephen J. Merrill
Ann V. LeFever

Marquette University

INTRODUCTION

The Humoral immune response to a foreign substance involves the production of antibody (Ab) specific for patterns (antigenic determinants) on that substance. If these determinants lie on a cell membrane as would happen if the substance is a bacteria, the binding of antibody to its specific antigen can activate the complement system, a collection of nine serum proteins along with at least one inhibitor and one inactivator. The activation of the system eventually results in cell lysis (destruction). This paper will discuss the recognition of the presence of antigen-antibody complexes (Ag∿Ab) and the resulting activation of the first components of the system. A model to the point of cell lysis is developed in [4] along with a more detailed derivation of this model.

The classical pathway of activation involves the activation of proteins which act as enzymes on other components. The form of the amplification phase is that of an enzyme cascase. Good references for this and similar systems are [1] and [2]. General references for the complement system are [5] and [6].

ISBN 0-12-045550-1

1. THE CLASSICAL PATHWAY

The components of the complement system, as well as the inactivators and inhibitors that control the system, are serum proteins. The classical pathway can be broken down into three functional units: 1) the recognition unit (C1 complex - C1q, C1r and C1s); 2) the amplification unit (C4, C2 and C3); and 3) the membrane attack unit (C5, C6, C7, C8 and C9). The membrane attack unit is the same for both the classical and alternative pathways of complement. Thus, the model of this unit also applies to the alternative pathway. The following flow diagram provides the basic scheme of the classical pathway of complement activation. (A bar over a component indicates activation; small case letters denote subcomponents.) See also figures 1-3.

$$C1q \xrightarrow{Ab} C1\bar{q} \xrightarrow{C1r} C1\bar{r} \xrightarrow{C1s} C1\bar{s} \xrightarrow{C4} C\overline{4b} + C4a$$

$$\searrow C2$$

$$\searrow C\overline{2a} + C2b \longrightarrow \sim$$

$$\longrightarrow C\overline{4b2a} \xrightarrow{C3} C\overline{4b2a3b} + C3a \xrightarrow{C5} C\overline{5b} + C5a$$

$$\text{(C3-convertase)} \qquad \text{(C5-convertase)}$$

$$\xrightarrow{C6} C\overline{5b6} \xrightarrow{C7} C\overline{5b67} \xrightarrow{C8} \xrightarrow{C9} C\overline{5b6789}$$

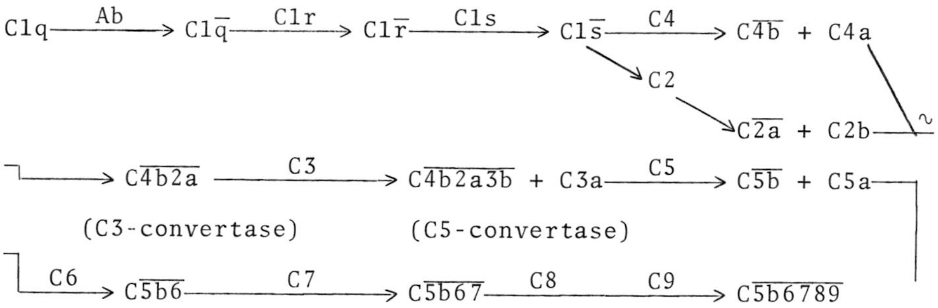

The recognition unit attaches to bound antibody and initiates the activation of the early components of complement required in the activation unit. An enzyme cascade, similar to those found in blood clotting systems [1], makes up the amplification unit. This unit amplifies the number of possible membrane attack units produced by the activation of a few initial complement components. Peptide chains of components in the membrane attack unit (C5, C7, C8 and C9) are thought to be

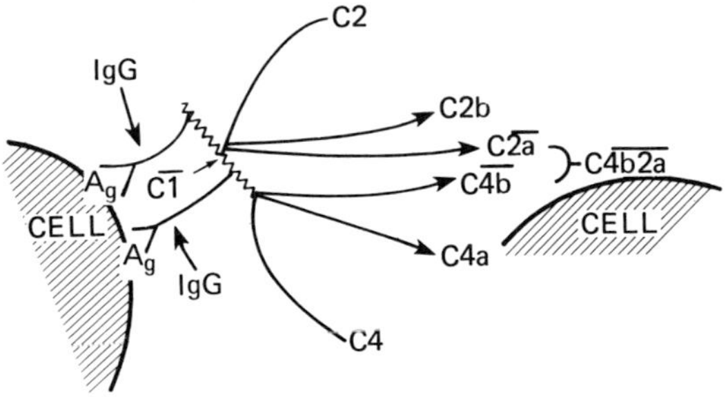

Recognition

Figure 1

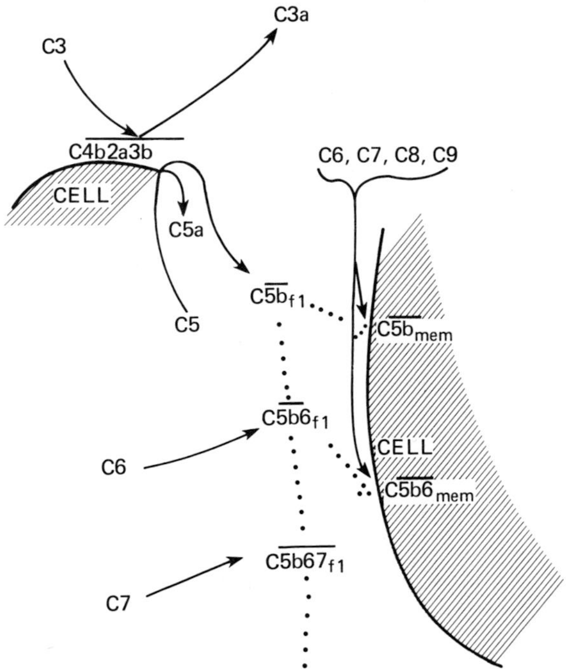

AMPLIFICATION

Figure 2

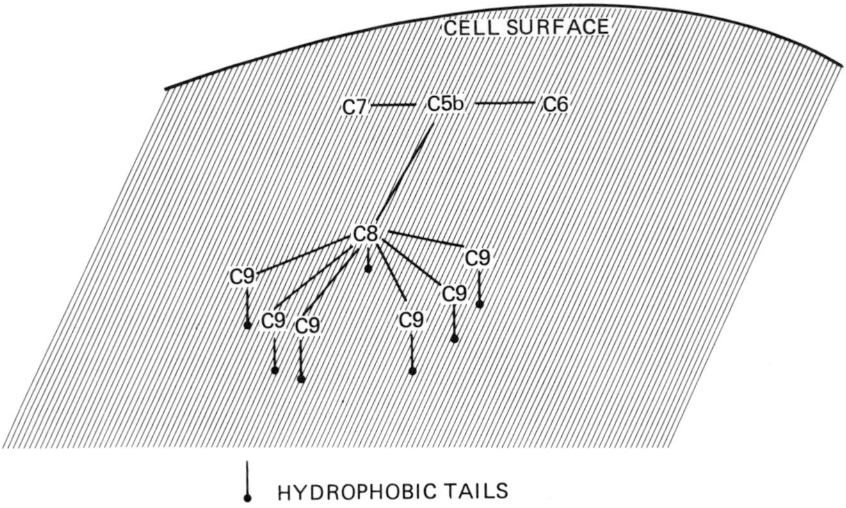

Membrane Attack

Figure 3

inserted cell membrane causing structural changes that result in hole or pore formation. These holes allow extracellular fluid to enter the cell and cause cell lysis.

There are a number of different mechanisms that control the amount of complement components produced and utilized in the lytic and pharmacologic activities of the complement system. These include: 1) the quantity of the components in serum; 2) the inhibitors and inactivators of the various components; 3) normal decay of components; and 4) rapid decay of components if they are not utilized rapidly in the subsequent complement sequence.

One of the goals of this model is to attempt to understand this complex system in light of those mechanisms responsible for its control.

Antibody and the complement system proteins are always present in serum and the question arises as to the cause of complement activation. The antibody molecule has been shown to undergo conformational changes when it binds to antigen or forms aggregates. Such changes may expose complement binding regions in the Fc portion of the antibody (C_H2 in the IgG molecule and C_H4 in the IgM molecule). Binding of the recognition unit to these antibody domains would then initiate the activation of the complement system.

The C1 protein is a complex of three proteins C1q, C1r and C1q. C1q binds to the antibody of the antigen-antibody complex resulting in activated C1q ($C1\overline{q}$) by a conformational change. Two molecules of antibody class IgG or one of class IgM is required to bind C1q [6]. As C1q changes conformation, C1r is activated to $C1\overline{r}$. The enzyme $C1\overline{r}$ cleaves C1s producing $C1\overline{s}$.

This is all taking place on one C1 molecule attached to one or two antigen-antibody complexes and the activation of C1s initiates the amplification phase. $C1\overline{r}$ and $C1\overline{s}$ are inhibited by the C1-Inhibitor ($C\overline{1}$-INH) which binds to the active sites (irreversibly).

The reactions are:

$$2(Ag \sim IgG) + C1q \underset{k_{-1}}{\overset{k_1}{\rightleftharpoons}} 2(Ag \sim IgG) \sim C1\overline{q}_G$$

$$Ag \sim IgM + C1q \underset{k^*_{-1}}{\overset{k^*_1}{\rightleftharpoons}} (Ag \sim IgM) \sim C1\overline{q}_M$$

$$C1r + C1\overline{q} \xrightarrow{k_2} C1\overline{r} + C1\overline{q}$$

$$C1r \xleftarrow{k_{-1}} C1\bar{r}$$

$$C1\bar{r} + C\bar{I} - INH \xrightarrow{k_3} C1\bar{r} \sim C\bar{I} - INH$$

$$C1s + C1\bar{r} \underset{k_{-4}}{\overset{k_4}{\rightleftarrows}} C1\bar{s} + C1\bar{r}$$

$$C1\bar{s} + C1 - INH \xrightarrow{k_5} C1\bar{s} \sim C1 - INH$$

Assuming the law mass action with [A] denoting concentration of A, $B = [C\bar{I} - INH]$, $B_1 = [C\bar{I} - INH \sim C1\bar{s}]$, $B_2 = [C\bar{I} - INH \sim C1\bar{s}]$, $CAb_M = [Ag \sim IgM]$, $CAb_G = [Ag \sim IgG]$ and C1 the constant concentration of C1, and assuming all rate constants positive, the model has

<u>equations</u>:

$$\frac{d[C1\bar{q}_M]}{dt} = k_1^*(C1-([C1\bar{q}_M] + [C1\bar{q}_G]))(CAb_M - [C1\bar{q}_M]) - k_{-1}^*[C1\bar{q}_M] \tag{1}$$

$$\frac{d[C1\bar{q}_G]}{dt} = k_1(C1-([C1\bar{q}_M] + [C1\bar{q}_G]))(CAb_G-2[C1q_G])^2 - k_{-1}[C1\bar{q}_G] \tag{2}$$

$$C1\bar{q} = [C1q_M] + [C1\bar{q}_G] \tag{3}$$

$$\frac{d[C1\bar{r}]}{dt} = k_2(C1\bar{q}-B_1-[C1\bar{r}])-k_{-1}\frac{[C1q_G]}{C1q}[C1\bar{r}] - k_{-1}^*\frac{[C1\bar{q}_M]}{C1\bar{q}}[C1\bar{r}]-k_3[C1\bar{r}](B-(B_1+B_2)) \tag{4}$$

$$\frac{d[C1\bar{s}]}{dt} = k_4([C1\bar{r}]-B_2-[C1\bar{s}])-k_{-4}[C1\bar{s}] - k_5[C1\bar{s}](B-(B_1+B_2)) \tag{5}$$

$$\frac{d\,B_1}{dt} = k_3[C1\bar{r}](B-(B_1+B_2)) \tag{6}$$

$$\frac{d\,B_2}{dt} = k_5[C1\bar{s}](B-(B_1+B_2)) \tag{7}$$

discussion:

(1) & (2). $C1\bar{q}$ production depends on C1 not activated ($C1 - ([C1q_M] + [C1\bar{q}_G]$ and the availability of antibody-antigen complex, either of type Ig_M ($CAb_M - [C1\bar{q}]$) or IgG where each $C1\bar{q}$ activated by IgG ($C1\bar{q}_G$) occupies two IgG-Antigen units. Both $C1\bar{q}_M$ and $C1\bar{q}_G$ can revert to C1q by disassociation with the antigen antibody complex.

(3). The concentration of activated $C1q = [C1\bar{q}_M] + [C1\bar{q}_G]$.

(4) C1r is activated when C1q is activated. $C1\bar{r}$ can become C1r when C1q disassociates and can be inhibited upon irreversible complexing with $C\bar{1}$-INH.

(5). C1s can be activated when the C1r on the same molecule has become activated. $C1\bar{s}$ can be inhibited or can revert to C1s.

(6) & (7). Concentration of C1-INH complexed with $C1\bar{r}$ and $C1\bar{s}$ respectively.

The initial conditions for (1)-(7) are from setting t = 0 in (I.C.) (t) below:

$$
\text{(I.C.)(t)}\quad
\begin{aligned}
&[C1\bar{q}_M](t) \geq 0\\
&[C1\bar{q}_G](t) \geq 0\\
&[C1\bar{q}_M](t) + [C1\bar{q}_G](t) \leq C1\\
&0 \leq [C1\bar{s}](t) \leq [C1\bar{r}](t) \leq [C1\bar{q}](t)\\
&B_1(t) \geq 0\\
&B_2(t) \geq 0\\
&B_1(t) + B_2(t) \leq B
\end{aligned}
$$

Behavior of (1)-(7). Definitions used below are as in [7].

Theorem 1. With initial conditions satisfying (I.C.)(0) solutions to (1)-(7) exist, are unique and satisfy (I.C.)(t) for all $t \geq 0$. This makes the region defined by (I.C.) (t), which is already compact, positively invariant. It then becomes a natural question to examine the asymptotic behavior of solutions in the (I.C.)(t) region.

Theorem 2. Within the region (I.C.)(t) there is a unique equilibrium (a_1, a_2) for (1) and (2). Moreover, any solution with initial condition satisfying (I.C.)(0), $[C1\bar{q}_M](t) \to a_1$, and $[C1\bar{q}_G](t) \to a_2$ as $t \to \infty$.

Now using theorems 1 and 2 together with the analysis of the asymptotically autonomous system (4) - (7) which results ([7]), one obtains:

Theorem 3. For initial conditions satisfying (I.C.)(0), solutions of (1)-(7) approach an equilibrium $(a_1, a_2, a_3, a_4, a_5, a_6, a_7)$ as $t \to \infty$. Moreover, if $a_3 = a_1 + a_2 < B$, $a_4 = a_5 = 0$ and if $a_3 > B$, $a_4 > 0$.

This theorem gives an important threshold behavior, as the asymptotic concentration of $C1\bar{s}$, a_5, will be zero if $a_3 < B$ and there will be positive $C1\bar{r}$ asymptotically if a_3 exceeds B. As a_3 depends directly on the concentration of antigen-antibody complexes, the inhibitor prevents activation if only a small antigen-antibody concentration is present.

Proofs of Theorems 1, 2, and 3.

Theorem 1. Existence and uniqueness for equations (1) and (2) follows immediately from standard theorems such as in [3]. Thus, $[C1\bar{q}_M](t)$ and $[C1\bar{q}_G](t)$ are defined on some internal of the form $(-\alpha, \alpha)$ for $\alpha > 0$. If $C1\bar{q}(t) = [C1\bar{q}_M](t) + [C1\bar{q}_G](t)$, note that both $\frac{[C1\bar{q}_M](t)}{C1\bar{q}(t)}$ and $\frac{[C1q_G](t)}{C1q(t)}$ are differentiable on $[0, \alpha]$ if $C1\bar{q}(0) \neq 0$ and on $(0,\alpha)$ if $C1\bar{q}(0) = 0$. In the latter case, both quotients can be extended continuously by L'Hospital's Rule:

$$\lim_{t \to 0} \frac{[C1\bar{q}_G](t)}{C1\bar{q}(t)} = \frac{k_1(CAb_G)^2}{k_1(CAb_G)^2 + k_1^*(CAb_M)}$$

and

$$\lim_{t \to 0} \frac{[C1\bar{q}_M](t)}{C1\bar{q}(t)} = \frac{k_1^*(CAb_M)}{k_1(CAb_G)^2 + k_1^*(CAb_M)}$$

Now examining (4)-(7), with the time dependent continuous functions $C1\bar{q}$, $\frac{[C1\bar{q}_G]}{C1\bar{q}}$ and $\frac{[C1\bar{q}_M]}{C1\bar{q}}$ on $(-\alpha, \alpha)$, we find that (4)-(7) satisfies the usual hypotheses sufficient for existence and uniqueness.

The solutions satisfy (I.C.)(t) for all time since the vector field defined in (1)-(7) always point into the region when examined on the boundary of that region. This is most easily seen by the use of standard comparison theorems. As any solution to (1)-(7) must at all times by tangent to the vector field, and the field is inward pointing, no solutions can leave that compact region. This also insures that the solutions can always be continued and this will exist for all $t > 0$.

Theorem 2. Setting $\frac{d[C1\bar{q}_M]}{dt} = 0$ and $\frac{d[C1\bar{q}_G]}{dt} = 0$ one finds

$$[C1\bar{q}_G] = C1 - [C1\bar{q}_M] - \frac{k^*_{-1}[C1\bar{q}_M]}{k^*_1(CAb_M - [C1\bar{q}_M])} \tag{8}$$

and

$$[C1\bar{q}_M] = C1 - [C1\bar{q}_G] - \frac{k_{-1}[C1\bar{q}_G]}{k_1(CAb_G - 2[C1\bar{q}_G])^2} \tag{9}$$

From (9), $[Cl\bar{q}_M] = Cl$ when $[Cl\bar{q}_G] = 0$ and $[Cl\bar{q}_M] = 0$ when $[Cl\bar{q}_G]$ is some number between 0 and $\min\{CAb_G, Cl\}$. Similarly from (8), $[Cl\bar{q}_G] = Cl$ if $[Cl\bar{q}_M] = 0$ and $[Cl\bar{q}_G] = 0$ for some value of $[Cl\bar{q}_M]$ between 0 and Min $\{CAb_M, Cl\}$. There must be at least one pair (a_1, a_2) such that if $[Cl\bar{q}_M] = a_1$ and $[Clq_G] = a_2$, both (8) and (9) are satisfied. Let (a_1, a_2) be the "first" such pair (with a_2 a minimum). We now show this point is unique. From (9),

$$\frac{\partial [Cl\bar{q}_M]}{\partial [Clq_G]} = -1 - \frac{k_{-1}}{k_1} \frac{[CAb_G \quad [Cl\bar{q}_G]}{(CAb_G - 2[Cl\bar{q}_G])^3} < -1$$

and from (8)

$$\frac{\partial [Cl\bar{q}_M]}{\partial [Cl\bar{q}_G]} = \frac{1}{\dfrac{\partial [Cl\bar{q}_G]}{\partial [Cl\bar{q}_M]}} = \frac{1}{-1 - \dfrac{k^*_{-1}}{k^*_1 (CAb_M - [Cl\bar{q}_M])}} > -1 .$$

By the above, the line $[Cl\bar{q}_M]\ a_1 = -1\ ([Cl\bar{q}_G] - a_2)$ is not crossed by either curve, (8) or (9), for $[Cl\bar{q}_G] > a_2$. As (a_1, a_2) was the first intersection, there can be no other, and this point is unique.

Examination of the phase plane for $[Cl\bar{q}_M](t)$ and $[Cl\bar{q}_G](t)$ (Figure 4) exposes the asymptotic behavior.

<u>Theorem 3</u>. <u>Definition</u> (Sell [7], page 37) Let $W \subseteq R$ and $f(x,t)$ be a continuous function from $W \times R \to R^n$ such that f is bounded and uniformly continuous on every set $K \times R^+$ where K is a compact subset of W and $R^+ = \{t \quad t \geq 0\}$. Then f is asymptotically <u>autonomous</u> iff there exists functions g and h in $C\ (WXR, R^n)$ such that

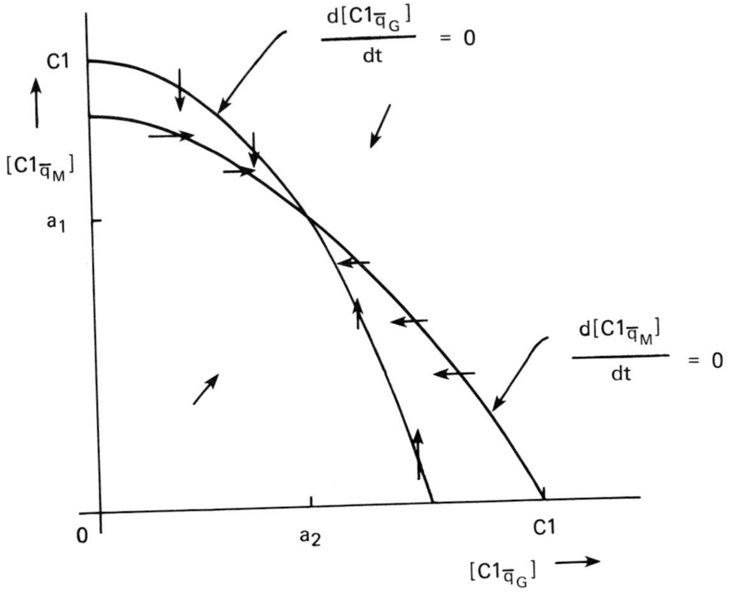

Phase Plane of (1)-(2)

Figure 4

(i) $f = g + h$

(ii) g is autonomous $(g(x,t_1) = g(x,t_2)$ for all $t_1, t_2 \in R$ and $X \in W)$

(iii) $h(x,t) \to 0$ as $t \to \infty$ uniformly on compact sets in W.

For our use here, W is the compact set defined by (I.C.)(t). From Theorem 2, $[C1\bar{q}_M] \to a_1$, $[C1\bar{q}_G] \to a_2$ and it then follows that the right hand sides of equations (4)-(7) are bounded and uniformly continuous on $W \times R^+$. With $h(x,t)$ defined in the natural way, (4)-(7) is asymptotically autonomous. It follows ([7], page 127) that the limiting behavior of (4)-(7) is identical to that of the autonomous $x' = g(x)$ given below:

$$\begin{cases} \dfrac{dx_1}{dt} = k_2(a_1 + a_2 - x_3 - x_1) - \dfrac{\underline{k}_1 a_2 + \underline{k}_1^* a_1}{a_1 + a_2} x_1 - k_3 x_1 (B - x_3 - x_4) \\ \dfrac{dx_2}{dt} = k_4(x_1 - x_4 - x_2) - \underline{k}_4 x_2 - k_5 x_2 \ (B - x_3 - x_4) \end{cases} \tag{10}$$

$$\begin{cases} \dfrac{dx_3}{dt} = k_3 x_1 \ (B - x_3 - x_4) \\ \dfrac{dx_4}{dt} = k_5 x_2 \ (B - x_3 - x_4) \end{cases} \tag{11}$$

As $x_3(t)$ and $x_4(t)$ are monotone increasing and bounded above, they too approach limits, a_6 and a_7. Thus, (10) is asymptotically autonomous, with limiting behavior that of:

$$\begin{cases} \dfrac{dy_1}{dt} = k_2(a_1 + a_2 - a_6 - y_1) - \left(\dfrac{k_1 a_2 + k_{-1}^* a_1}{a_1 \quad a_2} + k_3(B - a_6 - a_7) \right) y_1 \\ \dfrac{dy_2}{dt} = k_4 \ (y_1 - a_7 - y_2) - (\underline{k}_4 + k_5(B - a_6 - a_7)) y_2 \end{cases} \tag{12}$$

Now $y_1(t) \to \dfrac{k_2(a_1 + a_2 - a_6)}{k + k_3(B - a_6 - a_7) + (\underline{k}_1 a_2 + k_{-1}^* a_1)/(a_1 + a_2)} = a_4$

and $y_2(t) \to \dfrac{k_4(a_4 - a_7)}{k_4 + \underline{k}_4 + k_5(B - a_6 - a_7)} = a_5$

which must be the limits of $x_1(t)$ and $x_2(t)$ as $t \to \infty$. It then follows that as $t \to \infty$,

$$[C1\bar{r}](t) \to a_4$$

$$[C1\bar{s}](t) \to a_5 \ .$$

Moreover, if $a_1 + a_2 < B$, then since $\dfrac{d[C1\bar{r}]}{dt} \to 0$ and $\dfrac{d[C1\bar{s}]}{dt} \to 0$, $a_6 + a_4 \leq a_3$ and $a_7 + a_5 \leq a_4$. Together these imply that $a_6 + a_7 \leq a_3$. Thus, $B - a_6 - a_7 \geq B - a_3 > 0$.

To satisfy $\frac{dB_1}{dt} = 0$ and $\frac{dB_2}{dt} = 0$, one must have $a_4 = a_5 = 0$.

If $a_3 > B(\geq a_6)$, then $a_4 > 0$ as $\frac{d[C1\bar{r}]}{dt} \to 0$. Also, a_4 is an increasing function of a_3 and there is an unique value of a_3 (equal to $B + k_2 a_7/\underline{k}_1$ if $\underline{k}_1 = \underline{k}_1^*$) such that $a_4 = a_6$ and for a_3 larger, $a_5 > 0$.

REFERENCES

[1] Banks, H. T., "Modeling and Control in Biomedical Systems," Springer, 1975.

[2] Day, N. K., and R. A. Good, (eds.), "Biological Amplification Systems in Immunology," Plenum Press (1977).

[3] Hale, J., "Ordinary Differential Equations," Wiley-Inter-Science, 1969.

[4] LeFever, A. V., and S. J. Merrill, *The classical pathway of the complement system: A mathematical model*, (in preparation)

[5] Mayer, M. M., *Complement and complement fixation*, in Experimental Immunochemistry, Kabat, E. A. and Mayer, M. M. (ed.) (1964), pp. 133-240.

[6] Peltier, A., *Complement*, Immunology, J. F. Bach., (ed.), John Wiley and Sons (1978), pp. 220-247.

[7] Sell, G., "Topological Dynamics and Ordinary Differential Equations," Van Nostrand Reinhold, 1971.

SOLVABILITY OF NONLINEAR ELLIPTIC BOUNDARY VALUE PROBLEMS USING A PRIORI ESTIMATES

R. Kent Nagle

University of South Florida

Karen Singkofer

University of Southern California

INTRODUCTION

We are concerned with the existence of solutions of boundary value problems for nonlinear elliptic partial differential equations of the form

$$\begin{aligned} &Lu(x) + g(D^{\alpha}u(x)) = f(x) \qquad x \text{ in } \Omega \\ &\text{and} \\ &B_k u = 0 \qquad \text{on } \partial\Omega,\ k = 0, 1, \ldots, m-1 \end{aligned} \tag{1}$$

where $f \in L^{\infty}(\Omega)$ and the linear boundary value problem

$$\begin{aligned} &Lu = 0 \qquad \text{in } \Omega \\ &B_k u = 0 \qquad \text{on } \partial\Omega,\ k = 0, 1, \ldots, m-1 \end{aligned}$$

as a nontrivial set of of solutions spanned by a positive function θ. Let L be a uniformly elliptic selfadjoint differential operator of order $2m$ and $|\alpha| \le m$. Take $g : R \to R$ to be a continuous and bounded function and Ω to be a bounded domain in R^n with regular boundary $\partial\Omega$.

ISBN 0-12-045550-1

In recent years the special case when $|\alpha| = 0$, i.e. $Lu + g(u) = f$, has been studied extensively with much of the interest due to the paper by Landesman and Lazer [10]. The present emphasis is on using the asymptotic properties of g to obtain sufficient conditions on f in order for equation (1) to have a solution. These sufficient conditions are often referred to as Landesman-Lazer conditions.

In this paper we will show that often one must consider the local behavior of g in order to obtain sufficient conditions on f for equation (1) to have a solution. We will use the approach developed by Cesari [4,5] of combining nonlinear functional analysis and alternative methods to reduce equation (1) to a simpler one dimensional equation which can then be studied by considering the local structure of g. Our approach depends upon a priori estimates for the solution. Moreover, sufficient conditions for the existence of multiple solutions can be determined.

As references for the literature dealing with the Landesman-Lazer conditions, we refer the reader to the papers [2,3,5,6,8,11,16,18,19]. The case when the Landesman-Lazer conditions do not imply existence has been studied by [2,7,9, 13,14,15], of note are the papers [7,13,14,15] which do not rely upon the asymptotic properties of g.

In Section 2 we give our basic assumptions and reduce equation (1) to a one dimensional equation. The a priori estimates are also given in Section 2. Section 3 contains our existence and multiplicity results. In Section 4 we indicate how the results in this paper can be extended. Finally, in

Section 5 we discuss the need for considering the local properties of g instead of just the asymptotic properties. We also indicate directions for further research.

PRELIMINARY LEMMAS

For Ω a bounded domain in R^n let $C(\bar{\Omega})$, $C^\infty(\Omega)$, $C(\Omega)$, $L^p(\Omega)$ be the usual function spaces. For any integer $k > 0$, let $W^{k,p}(\Omega)$ be the Sobolev space of real functions in $L^p(\Omega)$ for which, in the sense of distributions, $D^i u \in L^p(\Omega)$ for $|i| \leq k$, with norm

$$\|u\|_{k,p} = \left\{\sum_{|i| \leq k} \int_\Omega |D^i u|^p \; dx\right\}^{1/p}.$$

Let $L = \sum_{\substack{|i| \leq m \\ |j| \leq m}} (-1)^i D^i(a_{ij} D^j)$ be a uniformly elliptic selfadjoint differential operator with $a_{ij} = a_{ji} \in C^\infty(\bar{\Omega})$ for $|i|$, $|j| \leq m$. The system of boundary operators associated with L, B_k, $k = 0, \ldots, m-1$ are assumed to have coefficients in $C^\infty(\partial\Omega)$ and satisfy complementary conditions with respect to L on $\partial\Omega$ in the sense of [1, Chapter 1].

Let $C_B^\infty(\Omega)$ denote the set of C^∞ function on $\bar{\Omega}$ satisfying $B_k u = 0$ on $\partial\Omega$ for $k = 0, \ldots, m-1$. Let V be the closure of $C_B^\infty(\Omega)$ in $W^{m,2}(\Omega)$. We will think of L as being generated via the Lax-Milgram Theorem by the bilinear form

$$B(u,v) = \sum_{\substack{|i| \leq m \\ |j| \leq m}} \int_\Omega a_{ij}(x) D^i u(x) D^j v(x) \; dx.$$

Our assumptions on L imply that B is a bounded, coercive bilinear form over V.

If we let P be the orthogonal projection onto the kernel of L in $L^2(\Omega)$, then from standard elliptic regularity results it follows that kernel of L = span of θ is contained in $C^\infty(\Omega)$ and $\theta \in W^{m,p}(\Omega)$ for all $m \geq 0$, $p \geq 1$. We can now uniquely express functions $u \in W^{2,p}(\Omega)$ by $u = v + c\theta$ where c is a constant and v is in $(I - P)\, W^{m,p}(\Omega)$.

On B we assume

(B) There are constants $\alpha_m > 0$ and $\mu \geq 0$ such that

$$B(u,u) \geq \alpha_m \|u\|_{m,2}^2 + \mu\|u\|_{0,2}^2$$

for all $u \in (I - P)V \subset (I - P)W^{m,2}(\Omega)$.

While assumption (B) is somewhat restrictive, bilinear forms generated by a uniformly elliptic differential operator satisfy such an inequality off the kernel of L. A further discussion of assumption (B) is given in [12, Lemma 2], [17, Lemma 1], and [14, Assumption (B)].

The nonlinearity $g(D^\alpha u)$, $|\alpha| \leq m$, is assumed to be a Nemytsky operator which is defined and hemicontinuous from $W^{m,2}(\Omega)$ into $L^2(\Omega)$. As a function $g : R \to R$ is taken to be continuous and bounded. Finally, we assume

(N) There exists constants δ, η with $\delta < \alpha_m$ and $\eta \leq \mu$ such that for all $u,v \in V \subset W^{m,2}(\Omega)$

$$(g(D^\alpha u) - g(D^\alpha v), u - v) \geq - \delta\|u - v\|_{m,2}^2 - \eta\|u - v\|_{0,2}^2.$$

The constants in assumption (N) can often be determined from the Lipschitz constants for g under an appropriate definition of Lipschitz. Moreover, (N) is a weaker assumption than monotonicity.

Using the projection operators P and $(I - P)$, equation (1) can be shown to be equivalent to the system

$$Lv + (I - P)g(D^{\alpha}(v + c\theta)) = f_1 \tag{2}$$

$$Pg(D^{\alpha}(v + c\theta)) = f_2 \tag{3}$$

where $f_1 = (I - P)f$ and $f_2 = Pf$.

Using results from monotone operator theory it has been shown in [15, Lemma 1 and Theorem 1] (Also [12], [14]) that

<u>Lemma 1</u>: Let L, P, and g satisfy the above assumptions. Then equation (1) is equivalent to

$$Pg(D^{\alpha}v(c) + D^{\alpha}c\theta) = f_2 \tag{4}$$

where $v(c)$ is the unique solution to equation (2) for a given constant c, $v(\cdot) : R \to (I - P)W^{m,2}(\Omega)$ is continuous, and $\|v(c)\|_{m,2} \leq (\alpha_m - \delta)^{-1}\{\|g(D^{\alpha}c\theta\|_{0,2} + \|f_1\|_{0,2}\}$.

Using our estimate of $\|v(c)\|_{m,2}$ and an argument used by Fucik and Krbec [7, Lemma 3] we can obtain an an priori estimate for $\|D^{\alpha}v(c)\|_{C(\overline{\Omega})}$. A proof when $|\alpha| = 0$ can be found in [14] and for $|\alpha| \leq m$ in [15].

<u>Lemma 2</u>: Let $v(c)$ be the unique solution to equation (2) as indicated in Theorem 1. Then there exists a constant $R(f_1)$ depending upon the bound for g, α_m, δ, α, and $\|f_1\|_{\infty,2}$ such that

$$\|D^{\alpha}v(c)\|_{C(\overline{\Omega})} \leq R(f_1) \tag{5}$$

for all real real numbers c.

EXISTENCE AND MULTIPLICITY RESULTS

In this section we give three existence theorems which illustrate the type of results which can be obtained by considering the local structure of the nonlinearity $g(D^{\alpha}u)$.

Theorem 1 considers the case when the kernel of L consists of the constants and $|\alpha| = 0$, i.e. $g(u)$. As a corollary we obtain sufficient conditions for multiple solutions to equation (1). Nonlinearities of the form $g(u) = A\sin Bu$, $g(u) = A(1 + u^2)^{-1}\sin Bu$, or $g(u) = Au(1 + u^2)^{-1}$ can be studied using Theorem 1.

Theorem 1: Let the conditions of Lemma 1 be satisfied with $|\alpha| = 0$ and $\theta \equiv 1$. If there exists constants c_1 and c_2 such that $\underline{g}(c_1) = \text{Min}\{g(c) : |c - c_1| \le R(f_1)\} > 0$ and $\overline{g}(c_2) = \text{Max}\{g(c) : |c - c_2| \le R(f_1)\} < 0$ where $R(f_1)$ is the bound given in Lemma 2, then there exists a constant $\rho > 0$ such that equation (1) has a solution provided $|\int_\Omega f_2| \le \rho$.

Proof: For $\theta \equiv 1$ equation (4) takes the form $\int_\Omega g(v(c) + c) = \int_\Omega f_2$. Define $\Gamma : R \to R$ by $\Gamma(c) = \int_\Omega g(v(c) + c) - \int_\Omega f_2$. Using inequality (5) and our assumption on $\underline{g}(c_1)$ we have $\Gamma(c_1) \ge \underline{g}(c_1)$ meas $(\Omega) - \int_\Omega f_2$. So for $\int_\Omega f_2 \le \underline{g}(c_1)$ meas (Ω), we have $\Gamma(c_1) \ge 0$.

In a similar fashion we can show that $\Gamma(c_2) \le 0$ if $\int_\Omega f_2 \ge \overline{g}(c_2)$ meas (Ω). Now select $\rho > 0$ so that $|\int_\Omega f_2| \le \rho$ implies $\Gamma(c_1) \ge 0$ and $\Gamma(c_2) \le 0$. Now by the continuity of g and $v(c)$ we get the continuity of $\Gamma(c)$ and hence there is a c_0 between c_1 and c_2 such that $\Gamma(c_0) = 0$. Since $\Gamma(c)$ is equivalent to equation (4), the theorem follows from Lemma 1. QED

Corollary 1: Let the condition of Lemma 1 be satisfied with $|\alpha| = 0$ and $\theta \equiv 1$. If there exists constants $c_1 < c_2 < \ldots < c_{n+1}$ such that $\underline{g}(c) > 0$ for i odd and

$\overline{g}(c_i) < 0$ for i even, then there exists a constant $\rho > 0$ such that equation (1) has at least n distinct solutions provided $|\int_\Omega f_2| \le \rho$.

Proof: In the proof of Theorem 1 choose ρ so that $\Gamma(c_i) > 0$ for i odd and $\Gamma(c_i) < 0$ for i even. It now follows that there exists constants c_i^*, $i = 1, \ldots, n$ between c_i and c_{i+1} such that $\Gamma(c_i^*) = 0$. Since distinct c_i^* give distinct solutions of the form $c_i^* + v(c_i^*)$, the corollary follows from Lemma 1. QED

Theorem 2 considers the more complicated case where g is a function of $D^\alpha u$. Here we must assume that $D^\alpha\theta$ is strictly positive, i.e. there are positive constants a and b such that $a \le D^\alpha\theta(x) \le b$ for all $x \in \Omega$. For simplicity we will assume θ is a positive function. Again the same nonlinearities $g(D^\alpha u) = A \operatorname{Sin} BD^\alpha u$, etc. can be studied. Previously nonlinearities involving the derivatives were only studied by a few authors [7,12,16,18] and then only using asymptotic methods.

Theorem 2: Let the conditions of Lemma 1 be satisfied with $D^\alpha\theta$ strictly positive and θ positive as indicated above. If there exists constants c_1 and c_2 such that

$$\underline{g}(c_1) = \begin{cases} \inf g(s) \quad \text{for} \quad ca - R(f_1) \le s \le cb + R(f_1) \\ \qquad\qquad \text{when} \quad c \ge o \\ \inf g(s) \quad \text{for} \quad cb - R(f_1) \le s \le ca + R(f_1) \\ \qquad\qquad \text{when} \quad c \le o \end{cases}$$

is positive and

$$\overline{g}(c_2) = \begin{cases} \inf\ g(s) \text{ for } ca - R(f_1) \le s \le cb + R(f_1) \\ \qquad \text{when } c \ge o \\ \inf\ g(s) \text{ for } cb - R(f_1) \le s \le ca + R(f_1) \\ \qquad \text{when } c \le o \end{cases}$$

is negative, then there exists a constant $\rho > o$ such that equation (1) has a solution provided $|\int_\Omega f_2\theta| \le \rho$.

Corollary 2: Let the conditions of Lemma 1 be satisfied with $D^\alpha\theta$ strictly positive and θ positive. If there exists constants $c_1 < c_2 < \ldots < c_{n+1}$ such $\underline{g}(c_i) > o$ for i odd and $\overline{g}(c_i) < o$ for i even, then there exists a constant $\rho > o$ such that equation (1) has at least n distinct solutions provided $|\int_\Omega f_2\theta| \le \rho$.

The proofs of Theorem 2 and Corollary 2 are essentially the same as the proofs of Theorem 1 and Corollary 1 with the obvious changes. (See [14 and [15]).

Finally Theorem 3 indicates the type of abstract theorems which can be proved by considering the local behavior of the nonlinearity. It is essentially a result due to Cesari [5, Theorem 34.i] where he states the result as if it is asymptotic in nature while in fact his proof is actually local in nature. Using Cesari's notation [5, Section 34] let E, P, and H be linear operators satisfying his assumptions (h_1), (h_2), and (h_3) on a Hilbert space X and let N be a nonlinear operator on X.

Theorem 3: Let there be a constant J_0 such that $\|Nx\| \le J_0$ for all $x \in X$ and assume (N_ε) there exists constants $K \ge J_0\|H\|$, $R_0 > 0$, $\varepsilon > 0$ such that $(Nx, x^*) \le -\varepsilon\|x^*\|$ (or $(Nx, x^*) \ge \varepsilon\|x^*\|$) for all $x \in X$, $x^* \in PX$, with $Px = x^*$, $R_0 \leq \|x^*\| < R_0 + J_0 + J_0^2/2\varepsilon$, $\|x - x^*\| \leq K$, then the equation $Ex = Nx$ has at least one solution.

Cesari assumes (N_ε) for $\|x^*\| \geq R_0$, however his proof only uses (N_ε) for $R_0 + J_0^2/2\varepsilon \leq \|x^*\| \leq R_0 + J_0^2/2\varepsilon + J_0$. In this theorem P is the projection onto the kernel of L, and there is no assumption that the kernel of L is one dimensional.

REMARKS

The method used in this paper can be used to obtain other similar results. (Also see [13, 14, 15, 16])

1. One can combine the local property that $\underline{g}(c_0) > 0$ Theorem 1 and an asymptotic argument with the limits $g(\pm\infty) = \lim_{s \to \pm\infty} g(s) = 0$ to show that for $0 \leq \int_\Omega f_2 \leq \rho$, equation (1) has at least two solutions, one corresponding to a $c_1 > c_0$ and the other corresponding to a $c_2 < c_0$. In particular, with the nonlinearity $g(u) = A(1 + u^2)^{-1}$ one would expect to have at least two solutions.

2. In Theorem 2, θ does not have to be positive. In general the conditions $\underline{g}(c_1)$ positive and $\overline{g}(c_2)$ negative would become

$$\underline{g}(c_1) \int_{\theta > 0} \theta \quad + \quad \overline{g}(c_1) \int_{\theta < 0} \theta \quad > \quad 0$$

and

$$\overline{g}(c_2) \int_{\theta > 0} \theta \quad + \quad \underline{g}(c_2) \int_{\theta < 0} \theta \quad < \quad 0 \, .$$

3. The assumptions in Theorem 1 and 2 concerning $\underline{g}(c)$ and $\overline{g}(c)$ can be reformulated in terms of the zeros of the function g. See the paper [14] for the case when g is a function of u.

4. The reduction to an alternative problem as given by Lemma 1 does not have to use the monotonicity properties of L and N. All that is needed is the conclusion of Lemma 1 remain valid, in particular, the existence of the continuous function v(c) with its bound in terms of the Sobolev norm.

5. Assumptions (B) and (N) can be weakened somewhat by projecting onto a larger subspace that the kernel of L. This method is used in the paper [15] to obtain a partial extension of Theorem 2.

6. The extension of the results to nonlinearities which depend upon more than one variable is routine, but tedious.

DISCUSSION

Since P is the orthogonal projection onto the kernel of L, equation (4) is equivalent to the equation

$$\Gamma(c) \overset{\text{def}}{=} \int_\Omega g(c + v(c))\ \theta = \int_\Omega f_2\ \theta\ .$$

It follows from Lemma 1 that $\Gamma(c)$ is a continuous function and hence, since g is bounded, $\Gamma(c)$ must be bounded. Therefore in order to determine the possible range of values for $\int_\Omega f_2\ \theta$ we need only determine the extreme values for $\Gamma(c)$. For most bounded functions g, these extreme values will not occur as c approaches either $\pm\infty$. For this reason we must consider the local properties of g.

It is interesting to observe that for any function f_1 we can find the v(c) given by Lemma 1. Moreover, v(c) is independent of f_2 and hence for a fixed f_1 there is a range of values for $\int_\Omega f_2\ \theta$.

Future research in this area could proceed in a number of directions:

1. A procedure for determining the maximum and minimum values of Γ is needed.

2. The existence of multiple solutions is determined by the oscillating behavior of g in connection with the estimates $R(f_1)$. More could be done along these lines.

3. Weaker conditions on L and g for which the conclusions of Lemmas 1 and 2 remain valid are needed.

4. When the kernel of L is multidimensional, Theorem 3 still remains valid. However, other than this abstract result, what type of local conditions on g will guarantee the existence of solutions to equation (1)? When do we have multiple solutions?

REFERENCES

[1] Agmon, S., A. Douglis and L. Nirenberg, *Estimates near the boundary for solutions of elliptic partial differential equations satisfying general boundary conditions I.*, Comm. Pure Appl. Math 12(1959), pp. 623-727.

[2] Ambrosetti, A., and G. Mancini, *Theorems of existence and multiplicity for nonlinear elliptic problems with noninvertible linear part,* Ann. Scuola. Norm. Sup. Pisa 5(1978), pp. 15-28.

[3] Brézis, H., and L. Nirenberg, *Characterizations of the ranges of some nonlinear operators and applications to boundary value problems,* Ann. Scuola. Norm. Sup. Pisa 5(1978), pp. 225-326.

[4] Cesari, L., *Functional analysis and periodic solutions of nonlinear differential equations*, "Contributions to Differential Equations," Wiley, New York, 1963, pp. 149-187.

[5] Cesari, L., *Functional analysis, nonlinear differential equations and the alternative method*, "Nonlinear Functional Analysis and Differential Equations", (Cesari, Kannan, Schurr, eds.), Marcel Dekker, New York, 1976, pp. 1-197.

[6] Fučík, S., *Nonlinear noncoercive boundary value problems*, "EQUADIFF IV Proceeding, Prague, 1977," (Jiří Fábèra, ed.), Springer Verlag Lecture Notes No.703(1979), pp. 99-109.

[7] Fučík, S., and M. Krbec, *Boundary value problems with bounded nonlinearity and general null-space of the linear part*. Math Z. 155(1977), pp. 129-138.

[8] Hess, P., *On semicoercive nonlinear problems*, Indiana Univ. Math. J. 23(1974), pp. 645-654.

[9] Hess, P., *Nonlinear perturbations of linear elliptic and parabolic problems at resonance: existence of multiple solutions*, Ann. Scuola. Norm. Sup. Pisa 5(1978), pp. 527-537.

[10] Landesman, E., and A. Lazer, *Nonlinear perturbation of linear boundary value problems at resonance*, J. Math. Mech. 19(1970), pp. 609-623.

[11] McKenna, P. J., and J. Rauch, *Strongly nonlinear perturbations of nonnegative boundary value problems with kernel*, J. Differential Equations 28(1978), pp. 253-265.

[12] Nagle, R. K., and K. Singkofer, *Equations with unbounded nonlinearities*, to appear in J. Nonlinear Analysis.

[13] Nagle, R. K., and K. Singkofer, *Nonlinear ordinary differential equations at resonance with slowly varying nonlinearities*, to appear.

[14] Nagle, R. K., and K. Singkofer, *Existence and multiplicity of solutions to nonlinear differential equation at resonance*, to appear.

[15] Nagle, R. K., and K. Singkofer, *Solvability of nonlinear elliptic equations at resonance via local and asymptotic estimates*, to appear.

[16] Nagle, R. K., K. Pothoven, and K. Singkofer, *Nonlinear elliptic equations at resonance where the nonlinearity depends essentially on the derivative*, to appear.

[17] Osborn, J., and D. Sather, *Alternative problems and monotonicity*, J. Differential Equations 18(1975), pp. 393-410.

[18] Shaw, H., *Nonlinear elliptic boundary value problems at resonance*, J. Differential Equations 26(1977), pp. 335-346.

[19] Williams, S., *A sharp sufficient condition for solution of nonlinear elliptic boundary value problems*, J. Differential Equations 8(1970), pp. 580-586.

ATTRACTORS IN GENERAL SYSTEMS[1]

Peter Seibert

Universidad Centro Occidental, Venezuela

INTRODUCTION

In V. I. Zubov's book [20], which appeared in 1957, Liapunov's second method was presented for the first time in a context more general than that of differential equations or related systems. In its initial chapters the notion of Liapunov function was adapted to dynamical systems in metric spaces, and in the last chapter so-called general systems (or dynamical polysystems, as they are also called) were considered. Since then, a theory of stability of dynamical systems and more general objects has been developed parallel to the more classical Liapunov theory; the results of stability theory for dynamical systems up to 1970 are presented in Bhatia's and Szegö's monograph [3].

Later, more abstract theories, in the spirit of the last chapter of Zubov's book, emerged. Some of these are of a topological flavor while others are of a more formal-logical nature. Both kinds were initiated by papers of D. Bushaw [5,6]. The first of these two points of view was subsequently developed independently by A. Pelczar (cf. [11,12], and the

[1]Este trabajo fué auspiciado por el CADIS.

ISBN 0-12-045550-1

literature cited there), and by our group, the starting point being the ideas described in J. Auslander's paper [1], centered around the concept of filter stability.

The aims of such a general, abstract approach include the simplification and unification of existing results and proofs, as well as finding new extensions and opening up of new fields of application by taking the theory out of the more special classical framework.

While in the classical second method, the study of Liapunov and asymptotic stability go together; in the framework of general systems, there are two separate theories, one concerning Liapunov stability, the other "asymptoticity," including attraction, which is found to be a more natural concept in this context than the traditional one of asymptotic stability. Eventually, both theories will be merged into a single, extended one which will cover both (and will include many basic concepts of classical analysis as special cases). A comprehensive exposition of the stability part of the theory (emphasizing connections with the classical theory) can be found in the report [18]. The present paper, for this reason, is entirely devoted to asymptoticity.

In the first two sections, the definitions of general flows and various concepts of asymptoticity and attraction are given. Section 3 contains a general Liapunov-type condition for asymptoticity and various more special results as corollaries. A special feature are the so-called uniform Liapunov functions which take care of a very general notion of uniform asymptoticity. Section 4 is concerned with inverse theorems, with emphasis on the uniform case. A Liapunov function does not always exist, and a useful criterion for its existence is still lacking.

In Section 5, Liapunov families are introduced which generalize those studied by Krasovskiǐ, Bhatia, Szegö, Yorke, and Salvadori. Here, also a fairly comprehensive inverse theorem is given. For a first advance into this area, see [7].

In Section 6, the question of existence of lower-semicontinuous Liapunov functions is considered. Due to results of Yorke, this is essential for the application of the general criteria. Finally, the connections between uniform attraction and asymptotic stability are touched upon briefly.

1. GENERAL FLOWS

1.0. Preliminaries. Let X be a non-empty set. $\mathcal{E}$ is a <u>collection</u>[2] on X if $\emptyset \neq \mathcal{E} \subset 2^X$. Given two collections $\mathcal{E}$ and $\mathcal{H}$ on X, we say

$\mathcal{E} \mathrel{(\!\!\prec} \mathcal{H}$ ("$\mathcal{E}$ is <u>dominated</u> by $\mathcal{H}$")

if every $E \in \mathcal{E}$ is contained in some $H \in \mathcal{H}$;

$\mathcal{E} \mathrel{-\!\!\!<} \mathcal{H}$ ("$\mathcal{E}$ is <u>coarser</u> than $\mathcal{H}$")

if every E contains an H.

The relations $(\!\!\prec$ and $-\!\!\!<$ are both transitive.

If $f : X \to 2^X$, we define for $E \subset X$,

$$fE = \cup \{f(x) \mid x \in E\},$$

and for a collection $\mathcal{E}$,

$$f\mathcal{E} = \{fE \mid E \in \mathcal{E}\}.$$

---

[2]We frequently denote collections by script capital letters, and their general elements by capital italics.

1.1 <u>Flows</u>. Let T be a linearly ordered set, which we shall call the <u>time scale</u>, denoting by t its elements, or <u>instants</u>. We associate to every $t \in T$ a function

$$f_t : X \to 2^X ,$$

calling $f_t(x)$ the <u>t-tail</u> of x. It is to be interpreted as the set of all points reachable from x at some instant $\geq t$. We introduce the following notations:

$$f_t \mathcal{E} := \{f_t E \mid E \in \mathcal{E}\} ;$$

$$f_T E := \{f_t E \mid t \in T\} ;$$

$$f_T \mathcal{E} := \{f_t E \mid t \in T , E \in \mathcal{E}\} .$$

If

$$t_1 \leq t_2 \text{ implies } f_{t_1} E \supset f_{t_2} E , \tag{1.1}$$

for all $E \subset X$, we call f_T a <u>flow</u> on X.

1.2. <u>Examples</u>. 1.2.1. Let (X,R,π) be a <u>dynamical system</u>[3] (with R denoting the real line). A flow in the sense just defined is given by the family f_T of functions f_t $(t \in R)$ defined by

$$f_t(x) = x[t,\infty) : = \{\pi(x,s) \mid s \geq t\} .$$

That it satisfies the condition (1.1) is a consequence of the group axiom. We call the sets $f_t(x)$ the <u>t-tails</u> of the orbit through x. When not specifying t, we will simply speak of a <u>tail</u> of the point x.

1.2.2. Consider a <u>non-autonomous differential equation</u> on a space Y, and denote by X the product space $R \times Y$, and by $\sigma(t;t_0,y_0)$ the solution satisfying the initial condition

---

[3] Our standard reference is [3].

$\sigma(t_0;t_0,y_0) = y_0$. A flow f_T is then defined on the space X by the functions

$$f_t(x_0) = \{\sigma(s;t_0,y_0) \mid s \geq t\}$$

$[x_0 := (t_0,y_0)]$. That this satisfies the condition (1.1) is obvious. Again, the f_t are the t-tails of the respective trajectories.

2. ASYMPTOTICITY

2.1. Basic definitions. By a system we mean a quadruplet $S = (X,f_T,\mathcal{U},\mathcal{A})$, where X is a set, f_T a flow on X, and $\mathcal{U}$ and $\mathcal{A}$ are collections on X. In particular, we will consider "special systems" where $\mathcal{U}$ consists of a single set U; in this case we write the system simply as $(X,f_T,U,\mathcal{A})$.

Definition 1. We say the system $(X,f_T,U,\mathcal{A})$ is asymptotic if

$$\mathcal{A} \prec f_T\, U .$$

This property can be described by saying that "every U is eventually contained in every $\mathcal{A}$."

Definition 2. The system $(X,f_T,\mathcal{U},\mathcal{A})$ is called asymptotic if every special system $(X,f_T,U,\mathcal{A})$, with $U \in \mathcal{U}$, is asymptotic; or

$$(\text{for each } U) \quad \mathcal{A} \prec f_T\, U$$

("every $\mathcal{A}$ contains a tail of every U").

Proposition 2.1. [7] Suppose $(X,f_T,\mathcal{U},\mathcal{A})$ is asymptotic. $\mathcal{A}' \prec \mathcal{A}$, and $\mathcal{U}' \in\prec \mathcal{U}$. Then $(X,f_T,\mathcal{U}',\mathcal{A}')$ is asymptotic.

The proof is straight forward.

2.2. Examples. 2.2.1. Suppose now that the flow f_T is defined on a topological space X, that A is a system of neighborhoods of a closed set $M \subset X$, and that U_o is a fixed neighborhood of M. Then, we say M is an A-attractor with a region of attraction containing U_o if the system $(X, f_T, \mathcal{U}_{U_o}, A)$, where $\mathcal{U}_{U_o} = \{\{x\} \mid x \in U_o\}$, is asymptotic, or, in other words, if every A contains a tail of every $x \in U_o$. This definition contains as special cases the concepts of attractor (both compact and non-compact) and semi-attractor of [3]. In these cases A is the metric neighborhood system (generated by the ε-neighborhoods) of M.

{We will call $\mathcal{U}_{U_o}$ the ultracollection on U_o, since any collection on U_o not containing the empty set is coarser than $\mathcal{U}_{U_o}$.}

2.2.2. Consider the same situation as in the preceding example, but now assuming X locally compact, metric, and M compact. We will denote by ν_x and ν_M the respective neighborhood systems of the point x and the set M. Then M is called a uniform attractor (or a strong attractor; [2]) with a region of attraction containing U_o if for every $V \in \nu_M$, and for every $x \in U_o$, there exist $V_x \in \nu_x$ and t such that $f_t V_x \subset V$.
(That this definition is equivalent to the one given in [3], is shown by proposition 1.2.3., p. 57, of that book.)

If $\mathcal{C}_{U_o}$ denotes the collection of all compact subsets of

U_o , and U_o is open[4], then the above definition of uniform attraction is equivalent to the asymptoticity of the system (X, f_T, C_{U_o}, ν_M) .

2.3. Uniform asymptoticity. We will now introduce a general notion of uniform asymptoticity which contains the one mentioned above as well as others.

Definition 3. Given a system $S_o = (X, f_T, U_o, A)$ and a cover $\mathcal{U}$ of U_o , then we call S_o $\mathcal{U}$-uniformly asymptotic if there exists a subcover $\mathcal{U}^* \subset \mathcal{U}$ of U_o such that $(X, f_T, \mathcal{U}^*, A)$ is asymptotic.

If, in the context of example 2.2.1. $\mathcal{U}$ is the ultracollection $\mathcal{U}_{U_o}$ on U_o , $\mathcal{U}$-uniform asymptoticity yields the attractor property.

If, on the other hand, in the context of example 2.2.2., $\mathcal{U}$ is the induced topology on U_o , $\mathcal{U}$-uniform asymptoticity yields uniform attraction, and the same is the case when $\mathcal{U} = C_{U_o}$ is chosen.

In general, a useful notion of attraction in the context of a flow on a topological space is obtained by choosing for $\mathcal{U}$ the topology, while A is a neighborhood system of the set M in question (usually the metric neighborhood system, supposing X endowed with a metric). This concept is much more practically meaningful than simple attraction, because the time constant t for a given neighborhood of M and point x is prevented from becoming arbitrarily large under arbitrarily small perturbations of x . In order to avoid confusion

[4]The region of attraction of a compact attractor in a locally compact metric space under a dynamical system is open (cf. [3], chapt. V).

with other concepts of uniform attraction, we will call this property <u>locally uniform attraction</u>, and, in the general case of a collection $\mathcal{A}$, we will speak of locally uniform asymptoticity. In the case of compact M, locally uniform attraction becomes uniform attraction as defined above.

<u>Definition 3a</u>. If, in definition 3, $\mathcal{U}$ is a topology on X, we will call the property described there, <u>locally uniform asymptoticity</u> (with respect to the topology $\mathcal{U}$)[5].

3. LIAPUNOV FUNCTIONS

3.1. Given a function $v : X \to [0,\infty] = \tilde{R}^+$, we define, for every $\beta > 0$, the sub-level set

$$L_v^\beta = \{x \in X \mid v(x) < \beta\} ,$$

and the collection

$$\mathcal{L}_v = \{L_v^\beta \mid \beta > 0\} .$$

<u>Definition 4</u>. v is an AL_2-<u>function</u>[6] for the system $S = (X, f_T, \mathcal{U}, \mathcal{A})$ if it satisfies the following conditions:

(AL_2.1) $\mathcal{A} \prec \mathcal{L}_v$;

(AL_2.2) $\mathcal{U} \,(\!\!\prec \mathcal{L}_v$;

(AL_2.3) (for every $\beta > 0$) $\mathcal{L}_v \prec f_T L_v^\beta$.

<u>Interpretation of the conditions</u>: (AL_2.1) means that v is bounded away from 0 on the complement of any A.

(AL_2.2) states that v is bounded on any U.

---

[5] The concept of locally uniform attraction was introduced by McCann in [10], who called it locally uniform asymptotic stability.

[6] A class of functions called AL_1 was studied in [7] and in item 12 of the bibliography of that paper.

(AL_2.3) means that, for any pair (α,β), $0 < \alpha < \beta$, the function v descends uniformly along the flow from the level β to the level α . {This condition is satisfied, for instance, if v admits a generalized total derivative, $\overset{*}{v}$, and satisfies a differential inequality of the form $\overset{*}{v} \leq -cv$, where c is a positive constant; or more generally, if $\overset{*}{v} < 0$, and bounded away from 0 outside of any L_v^β}.

Theorem 3.1. The system $S = (X, f_T, \mathcal{U}, \mathcal{A})$ is asymptotic if it admits an AL_2-function.

Proof. Given A and U , choose L_v^α and L_v^β such that $A \supset L_v^\alpha$, $U \subset L_v^\beta$, and then t such that $L_v^\alpha \supset f_t L_v^\beta$. The three conditions together yield $A \supset f_t U$, hence asymptoticity.

This result is essentially contained in Theorem 4, part i), case $j = 2$. pf [7], where it is stated for a family of functions, and without separation of the conditions (AL_2.1) and (AL_2.3) . It extends Pelczar's Theorem 4, [12], except that Pelczar admits Liapunov functions valued on more general sets which need not be linearly ordered.

3.2. Special cases. There are two essential ways of specializing the theorem: a) by specializing $\mathcal{A}$; b) by specializing $\mathcal{U}$. In all the usual cases, $\mathcal{A}$ is the metric neighborhood system, denoted in what follows by μ_M , of a closed set M (assuming X metric). Besides, M may or may not be assumed compact. In either case, condition (AL_2.1) reduces to

($AL_2$1') v is bounded away from 0 outside of any ε-neighborhood of M.

As far as the specialization of $\mathcal{U}$ is concerned, the three most important cases are the following:

3.2.1. $\mathcal{U}$ is a singleton, consisting of a fixed δ-neighborhood of M, which we denote by M_δ. Then M is a <u>uniform attractor</u> according to [3] (p.87), if $(X, f_T, M_\delta, \mu_M)$ is asymptotic. If M is compact and X locally compact, this definition coincides with the one given in [3] for that case (cf. 2.2.2, above), provided one does not want to specify the region of attraction. We, thus, have:

<u>Corollary 3.1.1. The closed set M is a uniform attractor under the flow f_T on the metric space X, if there exists a function v satisfying the conditions (AL_2.1') and (AL_2.3), and bounded on some δ-neighborhood of M.</u>

3.2.2. $\mathcal{U}$ is the ultracollection $\mathcal{U}_{U_o}$ on a neighborhood U_o of M. In this case asymptoticity of $(X, f_T, \mathcal{U}_{U_o}, \mu_M)$ gives the <u>attractor</u> property of M, with U_o as the region of attraction. {In the terminology of [3], M is called a semi-attractor, except when U_o is an ε-neighborhood of M, in which case it is called an attractor.}

<u>Corollary 3.1.2. The closed set M is an attractor under the flow f_T on the metric space X with the region of containing attraction U_o, if it admits a finite-valued function v satisfying the conditions (AL_2.1') and (AL_2.3).</u>

Indeed, condition (AL_2.2) in this case reduces to $u < \infty$.

3.2.3. $\mathcal{U}$ is the collection $\mathcal{C}_{U_o}$ of compact subsets of a neighborhood U_o of a compact subset M of the locally compact metric space X. In this case asymptoticity of $(X, f_T, \mathcal{C}_{U_o}, \mu_M)$ amounts to uniform attraction of M, and so we have:

Corollary 3.1.3. *The compact set M is a uniform attractor with its region of attraction containing U_o, under the flow f_T on the locally compact metric space X, if it admits a function v satisfying the conditions* $(AL_2.1')$ *and* $(AL_2.3)$ *and bounded on compact subsets of* U_o {(*in particular, if v is finite-valued and upper-semicontinuous*}.

These three corollaries fill a gap in [3], since no (simple) Liapunov-type conditions for attractors are given there. {A sufficient condition for a compact attractor in a dynamical system, involving a continuous function, is stated in the first theorem of sect. 16, [17].}

3.3. Uniform Liapunov functions. Let $U_o \subset X$, and let $\mathcal{U}$ be a cover of U_o. We call $v : X \to \tilde{R}^+$ a $\mathcal{U}$-*uniform* AL_2-*function* for the system (X, f_T, U_o, A), if it satisfies the conditions $(AL_2.1)$, $(AL_2.3)$, and

$(AL_2.2^*)$ There exists a subcover $\mathcal{U}^* \subset \mathcal{U}$ of U_o such that $\mathcal{U}^* \leftarrow\!< L_v$.

Corollary 3.1.4. *The system* (X, f_T, U_o, A) *is* $\mathcal{U}$-*uniformly asymptotic if it admits a* $\mathcal{U}$-*uniform* AL_2-*function*.

Indeed, let $\mathcal{U}^*$ be the subcover of $(AL_2.2^*)$. Then Theorem 3.1. yields asymptoticity of $(X, f_T, \mathcal{U}^*, A)$, which implies $\mathcal{U}$-uniform asymptoticity of the given system.

We now consider the case where $\mathcal{U}$ is a topology. Then, according to definition 3a, $\mathcal{U}$-uniform asymptoticity becomes locally uniform asymptoticity, and if A is the metric neighborhood system of a closed set M, this will be a locally uniform attractor. In this case, the last corollary becomes specialized as follows:

Corollary 3.1.5. The closed subset M of the metric space X is a locally uniform attractor for the flow f_T, with a region of attraction containing U_o, if there exists a function v satisfying the conditions (AL_2.1') and (AL_2.3), and bounded on some neighborhood of every point of U_o.

The last condition can be expressed appropriately by saying that "v is locally bounded on U_o." In particular, v is locally bounded if it is finite-valued and upper-semicontinuous.

4. THE EXISTENCE OF AL_2-FUNCTIONS

4.1. The existence of AL_2-functions requires a certain condition concerning the structure of the collections involved. In order to formulate these, we need the following definitions:

We call two collections E and H inner-equivalent if each is coarser than the other, outer-equivalent if each dominates the other, and bi-equivalent if they are both inner and outer equivalent.

A collection will be called inner-(outer)-admissible if it is inner (outer) equivalent to a decreasing (increasing) sequence of sets. If it is bi-equivalent to a countable nested collection of the order type of the integers, we call it bi-admissible.

Theorem 4.1. A necessary and sufficient condition for the existence of an AL_2-function for the system (X, f_T, U, A) is the existence of a bi-admissible collection R on X satisfying the following conditions:

($AL_2.\overline{1}$) $\mathcal{A} \prec \mathcal{R}$;

($AL_2.\overline{2}$) $\mathcal{U} (\!\prec \mathcal{R}$;

($AL_2.\overline{3}$) (for every $R \in \mathcal{R}$) $\mathcal{R} \prec f_T R$.

Proof. Necessity is immediate; indeed, if v as a function in question, L_v satisfies the three conditions and is bi-admissible.

Now suppose $\mathcal{R}$ satisfies the requirements of the theorem, and choose $\mathcal{B} = \{B_i \mid i \in Z\}$ bi-equivalent to $\mathcal{R}$ and increasing. Then define:

$v = e^i$ on $B_{i+1} - B_i$ ($i \in Z$, the set of the integers).

Obviously, (for every i) $L_v^{e^i} = B_i$. Now the verification of the conditions is trivial.

Examples where no AL_2-function exists. Let X be R^2 , $\mathcal{A}$ the (topological) neighborhood system of R , f_T and $\mathcal{U}$ unspecified. In this case the theorem excludes the existence of an AL_2-function because ($AL_2.\overline{1}$) cannot be fulfilled by any countable collection. {Of course, this is not one of the usual types of attractors, nor does it correspond to any type of behavior of the origin in a non-autonomous system of the usual kind, but it shows that the existence of a Liapunov function cannot be taken for granted in the presence of asymptoticity.}

4.2. In the case of uniform asymptoticity, it is easy to prove the existence of an AL_2-function. In order to include both the cases of compact and non-compact attractors (cf. 2.2.2, 3.2.1, resp.), we define the following concept which contains then both.

We say the "reduced system" $S' = (X, f_T, \mathcal{A})$ is <u>asymptotic</u> if

$$\text{(for every } A) \quad A \prec f_T A \ . \tag{4.1}$$

{Note that this is the same as asymptoticity of $(X, f_T, \mathcal{A}, \mathcal{A})$.}

By putting $\mathcal{A} = \mathcal{C}_{U_o}$, respectively $= \{V \in \mu_M \mid V \subset M_\delta\}$, we get the two concepts of uniform attraction à la Bhatia- Szegö for compact and non-compact sets.

We will now assume $T = R$ or Z, and suppose that f_T has a group structure $(f_{t_1} \circ f_{t_2} = f_{t_1 + t_2}, f_o = \text{id.})$.

Choose any $A_o \in \mathcal{A}$ and define: $\mathcal{R} = R_Z := f_Z A_o$. Given A, choose $i \in Z$ such that $A \supset f_i A_o = R_i$, implying $(AL_2.\overline{1})$; next, choose j such that $A_o \supset f_j A$, hence, $f_{-j} A_o = R_{-j} \supset f_o A = A$, proving $(AL_2.\overline{2})$; finally, given $i, j \in Z$, we have $R_i = f_i A_o = f_{i-j} \circ f_j A_o = f_{i-j} R_j$, hence, $(AL_2.3)$. Thus $\mathcal{R}$ has the properties specified in the theorem.

In summary, we can say that (with the qualifications stated) <u>asymptoticity of the "reduced system"</u> $(X, f_T, \mathcal{A})$ [i.e.(4.1)] <u>implies the existence of an</u> AL_2<u>-function; in particular, this is true in the case of uniform attractors</u>.

5. LIAPUNOV FAMILIES

5.1. Consider a family V of functions v from X into $\tilde{R}^+$. We introduce the following notations:

$$\hat{L}_v^\beta = \{x \in X \mid v \le \beta\}, \quad \hat{\mathcal{L}}_v = \hat{L}_v^\beta \mid \beta > 0\},$$

$$\hat{L}_V^\beta = \{\hat{L}_v^\beta \mid v \in V\}.$$

V will be called an $AL_2^!$-family of functions[7] for the system $S = (X, f_T, \mathcal{U}, \mathcal{A})$ if it satisfies the following conditions:

($AL_2^!$.1) $\mathcal{A} \prec \hat{L}_v^o$;

($AL_2^!$.2) (for every $v \in V$) $\mathcal{U} \Leftarrow\!\prec \hat{\mathcal{L}}_v$;

($AL_2^!$.3) (for every $v \in V$, $\beta > 0$) $\{\hat{L}_v^o\} \prec f_T \hat{L}_v^\beta$.

Theorem 5.1. S is asymptotic if it admits an $AL_2^!$-family of functions.

Proof. Let V be the family in question. Given A, U, choose v such that $A \supset \hat{L}_v^o$ (i.e., $v > 0$ outside of A), then β such that $U \subset \hat{L}_v^\beta$, and finally t such that $\hat{L}_v^o \supset f_t \hat{L}_v^\beta$. The three together yield $A \supset f_t U$, hence asymptoticity.

In order to formulate an inverse theorem, we assume $T = R$ and define:

a) The flow f_T is weakly t-continuous if for any $x, N \in \nu_x$, there exists a t such that

$$f^{[o,t)}(x) := f_o(x) - f_t(x) \subset N, \qquad (5.1)$$

and if also every "segment $f^{[\tau,t)}(x) := (f_\tau - f_t)(x)$ $(\tau < t)$ is nonempty. {In the case of a dynamical system, the relation (5.1) becomes $x[0,t) \subset N$.}

b) $\mathcal{A}$ is lower normal if $\mathcal{A} \prec \bar{\mathcal{A}} := \{\bar{A} \mid A \in \mathcal{A}\}$.

($\bar{\ }$: closure)

Theorem 5.2. If S is an asymptotic, weakly t-continuous system on a regular topological space, with $\mathcal{A}$ lower normal, then S admits an $AL_2^!$-family of functions.

---

[7]A similar family, called AL_2, was studied in [7]; see section 5.2.1, below.

Proof. Assign to every A the function

$$v_A(x) = \begin{cases} \inf \{t \mid f_t(x) \subset \bar{A}\} & \text{if } \{\ \} \neq \emptyset , \\ \infty \quad \text{if } \{\ \} = \emptyset , \end{cases}$$

and define $V = \{v_A \mid A \in \mathcal{A}\}$. We verify the three conditions:

($AL_2^!$.1): This means that outside of any $\bar{A}$, $v_A > 0$; indeed, if $x \notin \bar{A}$, choose $N \in \nu_x$ such that $N \cap \bar{A} = \emptyset$; then choose t such that $f^{[o,t)}(x) \subset N$, implying $f_\tau(x) \not\subset \bar{A}$, for $0 < \tau < t$, hence, $v_A(x) \geq \tau > 0$, Q.E.D.

($AL_2^!$.2): Given v, U, choose A such that $v = v_A$, then t such that $A \supset f_t U$, implying $v_A \leq t$ on U, i.e. $U \subset \hat{L}^t_{v_A}$ which proves the condition in question.

($AL_2^!$.3): Given v_A, $\beta > 0$, choose $y \in f_\beta \hat{L}^\beta_{v_A}$. Then we prove $v_A(y) = 0$, or equivalently,

(*) $f_\varepsilon(y) \subset \bar{A}$,

for any $\varepsilon > 0$. Select $x \in \hat{L}^\beta_{v_A}$ such that $y \in f_\beta(x)$. Then $v_A(x) \leq \beta$, hence, $f_{\beta+\varepsilon}(x) \subset \bar{A}$. Now,

$f_\varepsilon(y) \subset f_\varepsilon \circ f_\beta(x) = f_{\varepsilon+\beta}(x) \subset \bar{A}$, which proves (*).

5.2. Connection with other types of Liapunov families.

5.2.1. A class of families, called AL_2, very similar to our $AL_2^!$-families, was studied in [7]. The conditions ($AL_2^!$.1) and $AL_3^!$.3) were not separated. Also, in the proof, one function was associated to every U, and not, as in this paper, to every A. The latter has the advantage that in the case of inner-admissible $\mathcal{A}$, a countable family can be obtained. ($\mathcal{A}$ is more likely to have this property than $\mathcal{U}$). Moreover, the class AL_2 contains $AL_2^!$ as a special case.

5.2.2. Krasovskiǐ, in [8], Theorem 17.1., associates one function to each point of the region of attraction in the product space in order to formulate a sufficient condition for asymptotic stability of the null solution. This corresponds to our case of taking as $\mathcal{U}$ the ultracollection on the region of attraction. His functions present the same general characteristics reflected in our conditions ($AL_2^!.1$) through ($AL_2^!.3$), though with considerable differences in the details.

5.2.3. In [4], Bhatia, Szegö and Yorke formulate a necessary and sufficient condition for a global weak attractor in the case of an autonomous differential equation in R^n, associating to every member of a system of neighborhoods a lower-semicontinuous function. The construction of their functions is similar to ours, although the conditions are different. {The results are reproduced in [3], chapt. IX, sect. 2.}

5.2.4. In the context of a non-autonomous differential equation, Salvadori in [13,14] gives sufficient conditions for asymptotic stability assuming stability, by means of a family of functions, associating one function with every neighborhood. His conditions are more similar to ours, since he requires upper and lower estimates for the functions as well as an upper estimate for the (generalized) total derivative [corresponding to our condition ($AL_2^!.3$)].

6. THE QUESTION OF VERIFIABILITY OF THE CONDITIONS ($AL_2.3$) AND ($AL_2^!.3$)

6.1 The question of how non-increasing, resp. decreasing at a certain rate, of a non-Lipschitz Liapunov function in the context of a differential equation can be inferred from the behavior of its generalized total derivatives involving only the vector field and not the solutions, has been studied by Yorke in [19]. His principal result states that, whenever the function is lower-semicontinuous, it can be estimated along the trajectories without resorting to the solutions.

It is therefore essential for the application of Liapunov-type theorems that the functions involved can be assumed to be lower-semicontinuous.

6.2. We will first consider the problem in question in the case of theorem 3.1. For this, we need the following definition.

The flow f_T is called <u>weakly</u> x-<u>continuous</u> if, for any set $A \subset X$, and any t, $f_t\overline{A} \subset \overline{f_tA}$.

We define the <u>lower limit function</u> of a real-valued function v as

$$v_*(x) = \sup \{\inf v(N) \mid N \in \nu_x\}$$

For any function v, v_* is lower-semicontinuous.

One easily proves the following:

<u>Lemma</u>. For any real-valued function v and any $\beta > 0$, the equality $\hat{L}^{\beta}_{v_*} = \check{\bar{L}}^{\beta}_{v}$ holds.

<u>Theorem 6.1</u>. <u>If</u> v <u>is any</u> AL_2-<u>function for a given system</u> S, <u>then</u> v_* <u>is a lower-semicontinuous</u> AL_2-<u>function for the same system</u>, <u>provided</u> $\mathcal{A}$ <u>is lower normal</u>, <u>and</u> f_T <u>is weakly</u> x-<u>continuous</u>.

<u>Proof</u>. We show that v_* satisfies the three conditions for AL_2-functions:

(AL_2.1): Given A, choose $A' \in \mathcal{A}$ such that $A \supset \overline{A'}$. Then choose β such that $A' \supset L^{\beta}_{v}$. We have:

$$A \supset \bar{A}' \supset \overline{L_v^\beta} \supset \overline{\hat{L}_v^{\beta/2}} = \hat{L}_{v_*}^{\beta/2} \supset L_{v_*}^{\beta/2} \text{ , hence } A \prec L_{v_*} \text{ .}$$

(AL_2.2): Given U, choose β such that $U \subset L_v^\beta$, so that $\bar{\hat{L}}_v^\beta = \hat{L}_{v_*}^\beta \subset L_{v_*}^{2\beta}$, hence $U \leftarrow\prec L_{v_*}$.

(AL_2.3): Given α, $\beta > 0$, choose t such that $L_v^{\alpha/2} \supset f_t L_v^{2\beta}$. Then:

$$L_{v_*}^\alpha \supset \hat{L}_{v_*}^{\alpha/2} = \overline{\hat{L}_v^{\alpha/2}} \supset \overline{L_v^{\alpha/2}} \supset \overline{f_t L_v^{2\beta}} \supset f_t \overline{L_v^{2\beta}} \supset f_t \overline{\hat{L}_v^\beta} = f_t \hat{L}_v^\beta \supset f_t L_{v_*}^\beta,$$

hence $L_{v_*} \prec f_T L_{v_*}^\beta$, for any β

6.3. In the case of Theorem 5.2, we observe that the function v_A used in the proof is certainly lower-semicontinuous if f_T is a dynamical system. In general, this property amounts to a condition of upper-semicontinuity in x for each map f_t . Assuming this condition satisfied, the existence of a lower-semicontinuous Al_2'-family is guaranteed.

7. CONNECTIONS WITH ASYMPTOTIC STABILITY

The two concepts of asymptotic stability and uniform attraction are very intimately related. In the case of a compact attractor in a locally compact metric space, it is well known that both are equivalent (cf. [3], chapt. V).

In the more general case of a closed, positively invariant set in a metric space, one can show that locally uniform attraction implies asymptotic stability and is implied by "asymptotic uniform stability" (the stability being uniform, the attraction not necessarily)[8]. Similar theorems, involving also weak attractors, are presented [2], sect. 4.

REFERENCES

[1] Auslander, J., *On stability of closed sets in dynamical systems*, Sem. Diff. Eqs. Dynam. System, II, Univ. of Maryland, 1969, Lecture Notes Math., No. 144, Springer, (1970), pp. 1-4.

[2] Bhatia, N. P., *Attraction and nonsaddle sets in dynamical systems*, J. Diff. Equations 8(1970), pp. 229-249

[3] Bhatia, N. P., and G. P. Szegö, "Stability Theory of Dynamical Systems," Springer (1970).

[4] Bhatia, N. P., G. P. Stegö, and J. A. Yorke, *A Liapunov characterization of attractors*, Boll. Un. Mat. Ital., Ser. IV, 1(1969), pp. 222-228.

[5] Bushaw, D., *A stability criterion for general systems*, Math. Systems Theory, 1(1967), pp. 79-88.

[6] Bushaw, D., *Stabilities of Liapunov and Poisson types*, SIAM Review, 11(1969), pp. 214-225.

[7] Dankert, G., and P. Seibert, *Asymptoticity of general systems and Liapunov families*, Techn. Rep. DS 77-1, Dpto. Mat. Ci. Comp., No. 21, Univ. S. Bolivar, Caracas, Venezuela (1977), (to appera in Commentationes Mathematicae)

[8] Krasovskiǐ, N. N., " Stability of Motion, Stanford Univers. Press [Russion original: Moscow, 1959]. (1960).

---

[8] Here local uniform attraction must be understood in the sense of the property 1.2.3 of [3], p. 57, which is weaker than the definition given in section 2.3.

[9] LaSalle, J. P., *Some extensions of Liapunov's second method*, IRE Trans. Circuit Theory, Ct-7, (1960), pp. 520-527.

[10] McCann, R., *Embedding asymptotically stable dynamical systems into radial flows in* ℓ_2 , (to appear).

[11] Pelczar, A., *Stability of sets in Pseudo-dynamical systems*, III, Bull. Acad. Polon. Sci., Ser. Math. Astron. Phys., 20(1972), pp. 673-677.

[12] Pelcar, A., *La stabilite des eusembles daus des systemes pseudo-dynamiques locaux*, Zeszyty Nauk. Univ. Jagiellonsk., Prace Mat., 403(1975), pp. 7-11.

[13] Salvadori, L., *Sulla stabilita del motimento*, Le Matematiche (Sem. Mat. Univ. Catania) 24(1969), pp. 218-239.

[14] Salvadori, L., *Fimilie ad un parametro di funzioni di Liapunov Nello Studio della stabilita*, Sympos. Math. (Ist. Naz. Alta Mat.) 6(1971), pp. 309-330.

[15] Salzberg, P. M., and P. Seibert, *A unified theory of attraction in general systems*, Techn. Rep. DS 76-1, Dpt. Mat. Ci. Comp., No. 11, Univ. S. Bolivar, Caracas, Venezuela, (1976).

[16] Salzberg, P. M., and P. Seibert, *Asymptoticity in general systems*, Appl. Gen. Systems Research Proc. Nato Conf., Binghamton, N. Y., 1977; Plenum Pbl. Corp., (1978), pp. 371-379.

[17] Seibert, P., *Stability in dynamical systems*, Stab. Probls. of Sols. of Diff. Eqs., Proc. Nato Adv. Study Inst. Padua, Italy, 1965, Ediz. "Oderisi," Gubbio (Italy), (1966), pp. 73-94.

[18] Seibert, P., *Some recent developments in stability of general systems*, Appl. Nonlin. Analysis, Proc. Int. Conf., Univ. of Texas at Arlington, 1968, Academib Press (1979), pp. 351-371.

[19] Yorke, J. A., *Liapunov's second method and non-Lipschitz Liapunov functions*, Techn. Note BN-579, Inst. for Fluid Dyn. Appl. Math. Univ. of Maryland (1968).

[20] Zubov, V. I., *Methods of A. M. Liapunov and their Application*, Noordhoff, Groningen [Russian original: Leningrad, 1957]. (1964).

INFECTIOUS DISEASE IN A SPATIALLY HETEROGENEOUS ENVIRONMENT[1]

C. C. Travis
W. M. Post[2]
D. L. DeAngelis

Oak Ridge National Laboratory

INTRODUCTION

The incorporation of spatial heterogeneity into epidemic models can have two diametric effects. Models that incorporate continuous spatial distribution of individuals which interact strongly with neighbors and weakly with more distant individuals may demonstrate a damping effect resulting from geographic dispersion [1,4]. On the other hand, the fadeout of a disease in a subpopulation below the threshold size may be countered by reintroduction from other subpopulations [2]. Thus, interaction between subpopulations can effectively raise the infection rate so that the disease may persist in the total population even when it would fade out of each separate isolated subpopulation. It is toward this latter possibility that we focus our attention.

[1] Research supported by the National Science Foundation's Ecosystem Studies Program under Interagency Agreement No. DEB 77-25781 with the U. S. Department of Energy under contract W-7405-eng-26 with Union Carbide Corporation.

[2] Graduate Program in Ecology, University of Tennessee, Knoxville, Tennessee 37916.

ISBN 0-12-045550-1

Let us assume that the total population exists in a region of m population centers, where the ith center has a constant population, including susceptibles, infecteds, and immunes, of N_i. The members of each center make short visits to at least some of the other centers. To model this realistically, we could introduce complex model equations, but we shall attempt to simplify matters as much as possible so that our basic point can be clearly made. Actually, only certain fractions of each population will visit certain other population centers. However, we shall make the simplifying assumption that all members of each center spend the same amount of time visiting other centers, though the time spent visiting depends on the center. While visiting the other centers, infected visitors will have the chance of transmitting the disease to susceptibles in the visited center and susceptible visitors have a chance of acquiring the disease from infected members of the visited population center.

Define $x_i(t)$ as the number of susceptibles, $y_i(t)$ as the number of infecteds, and $z_i(t)$ as the number of immunes. Then the rate of change of each of these categories is assumed given by

$$\frac{dx_i}{dt} = a_i(x_i + y_i + z_i) - d_{xi}\, x_i - \sum_{j=1}^{m} b(T_{ij})\, x_i\, y_j \quad (1)$$

$$\frac{dy_i}{dt} = \sum_{j=i}^{m} b(T_{ij})\, x_i\, y_j - (d_{yi} + r_i) y_i \quad (2)$$

$$\frac{dz_i}{dt} = r_i\, y_i - d_{zi}\, z_i \quad (3)$$

The constants d_{xi}, d_{yi}, and d_{zi} are mortality rates, while r_i is the rate at which infected hosts recover, thereby joining the category of immunes. The constant a_i represents the reproductive rate, assumed, for simplicity, the same for each population category. The infectivity, $b(T_{ij})$, is expressed as a function of time T_{ij}, where $\sum_{i=1}^{m} T_{ij} = 1$. The expression $b(T_{ij})\ x_i\ y_j$ represents the rate of infection in subpopulation i resulting from contact with infected individuals of subpopulation center j. For simplicity, we further assume that the effect of the mortality rates, d_{xi}, d_{yi}, and d_{zi} is to keep the total population in the i-th center at a constant level, $x_i + y_i + z_i = N_i$.

We may thus restrict our attention to the equations

$$\frac{dx_i}{dt} = a_i\ N_i - d_{xi}\ x_i - \sum_{j=1}^{m} b(T_{ij})\ x_i\ y_j \tag{4}$$

$$\frac{dy_i}{dt} = \sum_{j=1}^{m} b(T_{ij})\ x_i\ y_j - (d_{yi} + r_i)\ y_i \tag{5}$$

We want to determine conditions under which interactions between subpopulations centers can cause a disease to become established when each center, in isolation, is incapable of supporting the disease. This is determined by examining local stability of the equilibrium point $\hat{N} = (N_1, N_2, \ldots, N_m, 0, 0, \ldots, 0)$. The perturbed equations obtained from (4) and (5) in the neighborhood of $\hat{N}$ are

$$\begin{bmatrix} \frac{dx}{dt} \\ \\ \frac{dy}{dt} \end{bmatrix} = \begin{bmatrix} A_{11} & A_{12} \\ A_{21} & A_{22} \end{bmatrix} \begin{bmatrix} x \\ y \end{bmatrix} \tag{6}$$

where dx/dt, dy/dt, x, and y are $(m \times 1)$ vectors and A_{11}, A_{12}, A_{21}, and A_{22} are $m \times m$ matrices of the form

$$A_{11} = \text{Diag } [-d_{xi}]$$

$$A_{12} = [-b(T_{ij})\ N_j]$$

$$A_{21} = [0]$$

$$A_{22} = \text{Diag } [-(d_{xi} + r_i)] + [b(T_{ij})\ N_j]$$

Since $A_{21} = 0$, the set of eigenvalues of A is the union of the eigenvalues of A_{11} and A_{22}. By inspection, the submatrix A_{11} contributes eigenvalues with negative real parts. If the equilibrium point $\hat{N}$ is to be unstable, then the submatrix A_{22} must have at least one eigenvalue with a positive real part. We have established the following theorem.

Theorem. A disease can become established in a spatially heterogeneous population if and only if the submatrix A_{22} of Eq. (6) has at least one eigenvalue with a positive real part.

The sign pattern of the matrix A_{22} permits the use of simple criteria to determine stability. To establish the criteria, we introduce the notion of an M - matrix (see Plemmons [5] for a thorough review).

Definition. A $k \times k$ matrix, $M = (m_{ij})$, $(1 \le i, j \le k)$ is said to be an M - matrix if $m_{ij} \le 0$ for all $i \ne j$ and if any one of the following equivalent statements is true;

(i) all the principal minors of M are positive;

(ii) all eigenvalues of M have positive real parts;

(iii) there is a vector $u > 0$ such that $M u > 0$;

(iv) there is a vector $v > 0$ such that $M^T v > 0$.

We see that all the eigenvalues of the matrix A_{22} have negative real parts of and only if $-A_{22}$ is an M - matrix. This establishes the following corollary to the theorem.

Corollary. A disease can become established in a spatially heterogeneous population if and only if $-A_{22}$ is not an M - matrix.

We shall now explore the relationship between the functional form of the rate of contagion and whether or not a disease can become established in a population. Suppose that infectivity is linearly proportional to the fraction of time two populations are in contact, that is $b(T_{ij}) = b\,T_{ij}$. Choosing $v = (1/N_1, 1/N_2, \ldots, 1/N_m)$, the vector $-A_{22}^T v$ consists of elements of the form

$$(d_{yi} + r_i)/N_i - \sum_{j=1}^{m} b\,T_{ji} \qquad (7)$$

Since $\sum_{j=1}^{m} T_{ji} = 1$, a sufficient condition that $-A_{22}$ be an M - matrix is

$$(d_{yi} + r_i) - b\,N_i > 0, \quad i = 1, 2, \ldots, m\,. \qquad (8)$$

These conditions are the well-known [1,3] threshold condition for determining when a disease cannot become established in an isolated population. Thus, in the case when infectivity is linearly proportional to the fraction of time subpopulation centers are in contact, the threshold condition for maintenance of a disease in an entire population is identical to the threshold condition for the maintenance of the disease in each of the isolated subpopulation centers. In other words, the disease cannot become endemic in the entire population unless it is endemic in some isolated subpopulation.

We now study in greater detail the relationships between the functional form of the rate of contagion and whether or not a disease can become established in a population. To simplify our analysis, consider the case of two population centers which cannot support a given disease in isolation. As we have seen above, in order for an epidemic to occur in this situation, contact between population centers must enhance infectivity. We introduce two new variables, Δ_1, Δ_2, which express the increase in the rate of infection a subpopulation experiences due to migration or contact with other subpopulation centers. The variables are defined as

$$\Delta_1 = b\ (T_{11}) + b\ (T_{21}) - b_0 \tag{8}$$

$$\Delta_2 = b\ (T_{12}) + b\ (T_{22}) - b_0 \tag{9}$$

Choosing model parameters and a particular form of the infectivity function, we can evaluate the effect of migration on the progress of the disease. Figure 1 summarizes the results of such a calculation for various values of b_0 and an infection rate function of the form $b(T) = 2\ b_0 T/(1 + T)$. Each subpopulation center was assumed to have the same population size $(N_1 = N_2 = 200)$ and the same rate of removal of infecteds from the infected class $(r_1 + dy_1 = r_2 + d_{y2} = 0.8)$. If the point specified by Δ_1 and Δ_2 lies above the hyperbola for the appropriate values of b_0, the disease will become endemic. If the point lies below the hyperbola it will not. The threshold condition necessary for each subpopulation center to support the disease in the absence of interaction is $b_0 > 0.004$. If b_0 is less than this critical rate, then the disease can become established only if there is sufficient interaction between subpopulation centers. This will occur when Δ_1 and Δ_2 are large enough for the point (Δ_1, Δ_2) to lie above the appropriate hyperbola.

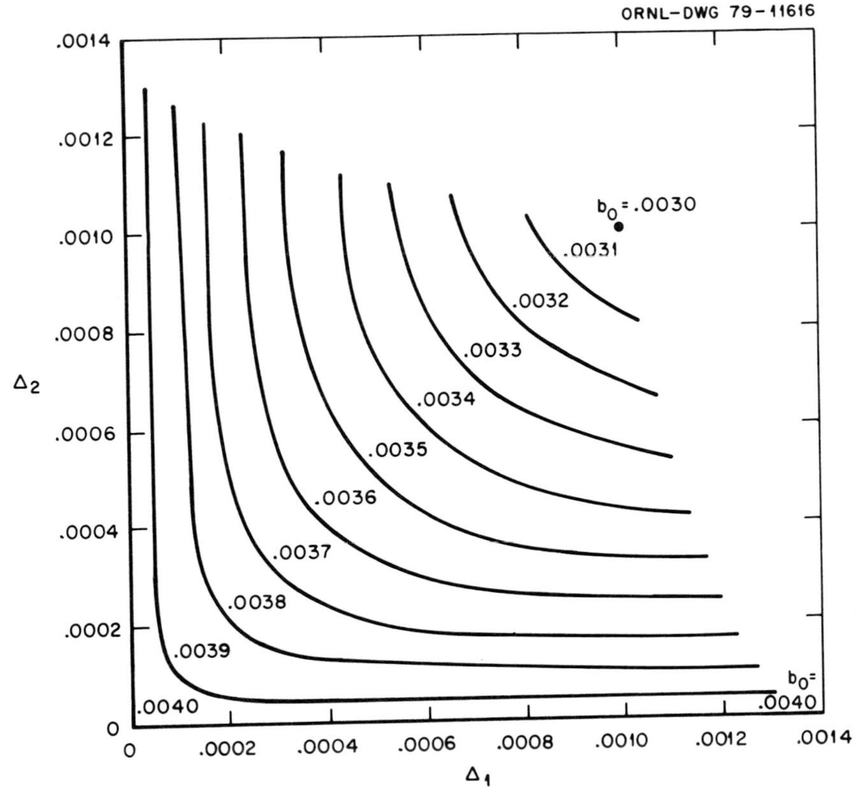

Figure 1

These curves define the enhancement in infectivity resulting from contact that is required for a disease to become endemic in a population of two interacting subpopulations. (See text for additional explanation.)

There are two other important facts that Figure 1 makes clear. First the shape of the curve separating the stable from the unstable region suggests that this transition occurs when the product Δ_1 Δ_2 exceeds some constant. This requires that both subpopulations must mix with each other for the disease to become established. Secondly, only a small section of each hyperbola is drawn. This is due to the fact that Δ_1 and Δ_2 have maximum values which depend on the infectivity function $b(t)$, and the magnitude of b_0. As a result, there is an infection rate (in the present case $b_0 = 0.003$), below which a disease cannot become established regardless of the degree of mixing between the subpopulations.

REFERENCES

[1] Bailey, N. T. J., *The Mathematical Theory of Infectious Diseases and Its Applications*, Hafner, New York, 1975.

[2] Black, F. L., "Measles Endemicity in Insular Populations: Critical Community Size and Its Evolutionary Implications," J. Theoret. Biol. 11, 207-211 (1966).

[3] Lewis, T. "Threshold Results in the Study of Shistosomiasis," Math. Biosci. 30, 205-211 (1976).

[4] Noble, J. V., "Geographic and Temporal Development of Plagues," Nature 250, 726-728 (1974).

[5] Plemmons, R. V., "M-matrix Characterization. I. Nonsingular M-matrices," Lin. Alg. and Appl. 18, 175-188 (1977).